Pipe & Excavation Contracting

by

Dave Roberts

Craftsman Book Company
6058 Corte Del Cedro, Carlsbad, CA 92009

The author wishes to express his appreciation to the following companies for furnishing materials used in the preparation of various portions of this book.

AGL Corporation — P.O. Box 189, Jacksonville, AR 72076

Allied Steel & Tractor Products, Inc. — 5800 Harper Road, Solon, OH 44139

Berger Instruments — 4 River Street, Boston, MA 02126

Clow Corporation — 2021 Midwest Road, Oak Brook, IL 60521

Deere & Company — John Deere Road, Moline, IL 61265

Elco International, Inc. — 111 Van Riper Avenue, Elmwood Park, NJ 07407

Griswold Machine & Engineering, Inc. — Hwy. M-60, Union City, MI 49094

Laser Alignment, Inc. — 2850 Thornhills, Grand Rapids, MI 49506

A.Y. McDonald Mfg. Co. — 12th & Pine Streets, Dubuque, IA 52001

McGraw-Hill Book Company — 1221 Avenue of the Americas, New York, NY 10020

Mueller Company — Decatur, IL 62525

Pacific Builder & Engineer — McKellar Publications, Inc., 410 W. Arden Ave. #203, Glendale, CA 91203

RayGo — 9401 85th Avenue North, Minneapolis, MN 55440

Library of Congress Cataloging-in-Publication Data

Roberts, Dave, 1951 —
    Pipe & excavation contracting / by Dave Roberts.
        p.   cm.
    Includes index.
    ISBN 0-934041-22-9
    1. Excavation.   2. Pipe lines.   I. Title.   II. Title: Pipe and excavation contracting.
TA730.R63 1987
624.1'52--dc19   87-24205
                                                                        CIP

CONTENTS

1

ESTIMATING, BIDDING, BONDING

Pipeline contracting can be a very rewarding business — if you know what you're doing and do it right. It's less plagued by the economic cycle that pushes many excavation and construction contractors to the brink of bankruptcy when the market for residential and commercial building dries up. And the profit margins are usually better because it's harder to get started in this business. You need money, manpower, skill and knowledge to get into sewer, water and drainage line contracting. That limits your competition and can fatten your margins.

Just owning a backhoe doesn't make you a pipeline contractor. There's far more to it than simply digging a hole and filling it up. If you've considered pipeline work but have never done any, keep reading. This book is for you. If you've bid some pipeline work but aren't satisfied with the results, the tips and suggestions in this manual may be just what you need to succeed where others are striking out.

True, every pipeline contractor (or utility contractor, as we're sometimes called) needs special skills, know-how and expensive equipment. But if you can estimate accurately, pay close attention to your costs, keep your overhead low, do nothing but quality work, and eliminate wasted time and materials, you've got a good chance of making it in this demanding business.

Understand from the beginning that Rome wasn't built in a day. No one jumps into this field and makes big money handling million dollar jobs right from the start. You're going to have to crawl before you walk and walk before you run. But if you've got the determination to work hard and are willing to take the time to study and understand what I've written here, you're well on your way to a good professional career.

Nearly all utility line work is awarded by competitive bid. If
your bid is too high, you won't get the job. If your bid is too
low, you'll get the job but lose your shirt. A few jobs like that
and your company will go belly up.

Every year, hundreds of men (and a few women) set out to
become underground utility line contractors. Many were making
good wages as tradesmen, equipment operators or supervisors in
the business, and decided to try their hand at bidding and
contracting for the work they did formerly as wage earners.
Unfortunately, few survive more than a few years in the
business.

But don't let the failure of others discourage you. The fact
that you're reading this book shows that you plan on going into
the business as an *informed* entrepreneur. And if you're already
in the business, it shows you plan on finding out what not to do
by *my* mistakes instead of yours. Work with care, both at your
estimating table, and on the jobsite. With the information in
this book, and, I can't deny, with a little bit of luck, you can
have a successful and profitable underground utility contracting
business. That's what I wrote this book for.

There's plenty of work available if you're willing to move, be
flexible, and not shy away from challenges. The private
developers of residential and commercial projects need utility
contractors. The government — the U.S. Forest Service, the
B.L.M., the Soil Conservation Service, military bases — all are
regular customers. And many government contracts require that
a percentage of the contract be let to small businesses or
minority-owned businesses.

While this book is primarily aimed at the underground utility
contractor, be aware that you can get a wider scope of work if
you're willing to compile bids and manage contracts that
contain other kinds of work, such as mechanical, electrical, and
building work. As a small contractor, you'll get a lot more
business if you're willing to take on contracts more complex
than a simple pipe job. Move into unfamiliar areas very
cautiously, however. Use reliable subcontractors and beware of
getting into fields that are really beyond your experience and
knowledge.

Starting at the Beginning
In this book, I'll describe all the common pitfalls and how to
avoid them. We'll take a detailed look at the practical know-

how you'll use on the job every day — estimating and bidding, equipment and crew performance, practical work methods and techniques, equipment operation, surveying, site clearance, compaction, water systems, and sewer systems. In short, I'll cover everything a utility line contractor needs to know to thrive in this business.

You'll need a working knowledge of the skills of the four important professionals on every utility line team: the pipe layer, the equipment operator, the foreman and the contractor. The success of your company depends on how well each of these four work together toward a common goal.

The pipe layer— Laying pipe can be fairly simple work. But it seldom is. Crosslines, groundwater, rocky soil, or a combination of all these will always complicate the work. Installing miles of pipe exactly right and at a fast pace is no easy task. Your pipe layers especially can make or break you. I'll explain this important task in detail later.

The equipment operator— Modern earth-moving equipment can be maneuvered as easily as your family car. But that isn't the whole story. The trick is to get *maximum productivity* from each man and machine hour. Many heavy equipment operators are careless or have only average skills. As an underground utility contractor, your survival depends on finding and keeping highly skilled, experienced, and motivated operators.

The foreman or superintendent— Another name for the foreman, superintendent or crew leader is *pusher.* A good pusher knows utility line construction backwards and forwards. He probably worked as a pipe layer and equipment operator before becoming a supervisor. It's much easier to supervise work you've done yourself. There's no substitute for practical experience.

Most of a superintendent's time is spent dealing with minor catastrophes so the job stays on schedule. Bad weather, machinery breakdowns, wrong materials deliveries — anything that can go wrong, and does, is his problem. And at the same time, he has to stay on good terms with his crew, the engineers, the inspectors, the owner, and the public — while he gets each to do what they should at the time and place they should be doing it.

In the final analysis, his chief job responsibility is making sure the project finishes on time and under budget. If it does, nothing he did along the way makes much difference. If it doesn't, all the excuses he can think of are irrelevant. A good superintendent is worth his (or her) weight in progress payments. If you have a good superintendent and can't keep him up to his eyeballs in work, send me his name and phone number. I'll take it from there.

The contractor— You need to know everything a good superintendent knows, plus lots more: taxes, insurance, handling personnel problems, borrowing and investing, estimating and bidding, hiring and working with reputable professionals, and then some. Your job is the biggest and most difficult of all.

By now you should see that underground utility contracting is demanding and risky work. All construction contracting is. But underground work is especially risky. And the highly competitive nature of utility line work doubles the risk for all players.

But the potential rewards are equal to the risk. Some utility line contractors have retired at a relatively early age with enough earnings and savings to support them comfortably for a lifetime. Many started from scratch, probably with less knowledge and no more cash than you have. Not everyone can do it, of course, but you'll never know how good you can be until you try. So my advice is: Give it a try.

Let this book be your first step toward a successful career in underground utility contracting. Or allow it to boost your existing career by showing techniques and methods that will increase your profitability.

This chapter will explain the best way to prepare an estimate and submit the bid. Once I've covered the procedure, I'll test your knowledge with a sample bid on an actual underground utility project. Later in this chapter there's a sample estimate for a small drainage and slide abatement project.

Bonding is very important in the utility line contracting business — far more important than in home building or remodeling, for example. If you can't get bonded, you're closed out of most of the larger, more profitable jobs. That's why there's an extensive section in this chapter on getting bonding and increasing your bonding capacity.

Nature of the Business

As an underground utility contractor, most work will come
from either private developers or publicly-funded municipal
projects.

Land developers usually let subcontracts for street
construction and utility installation. Some branch of city
government will require that your work meet their standards.
They'll inspect what you've done. That just makes sense. When
you're finished, the city will accept responsibility for
maintaining what you install.

Public projects include work like installing new utilities and
replacing old utilities in existing cities or developments. This
type of work is funded by selling bonds to the public, or by
federal loans and grants. A small city, for example, may borrow
federal money to pay for a replacement water or sewer system.
This money will be repaid by water users over a period of years.

It's in the owner's best interest to get bids from several
contractors. And on publicly-funded jobs, no qualified
contractor can be excluded. You'll see upcoming projects
advertised in newspapers and listed in local and regional bid call
magazines. Some magazines and newsletters list projects being
put out for bid and report the prices submitted by successful
bidders.

Some large projects are divided into several small projects
because smaller jobs attract more bidders. Most contractors
can't handle a $2,000,000 project. So don't be surprised to see a
$2,000,000 job divided into four $500,000 contracts. The more
bidders there are, the stiffer your competition will be.

The Design Engineer

Whether the job you're bidding is public or private work, the
project usually begins with a utility line design and an
engineering company. On some large jobs the contractor's
engineering team prepares the design drawings, oversees the
work, authorizes progress payments, and issues a certificate
upon completion. The design and build contracts are usually
undertaken by large multinational companies.

On most jobs, a separate engineering company is hired by the
owner. The engineering company usually works for a percentage
of the projected cost, or a fixed fee. On other jobs, the design

work will be done by an engineering firm hired by the city, county or developer, or by engineers employed by the contracting authority.

Following current engineering standards, the designer draws up plans and specs that show the materials and workmanship required. Once he's designed the project, he'll make the *engineer's estimate.* This is the first real cost estimate on the work to be done. But it probably won't be based on a careful analysis of labor, material and equipment costs. Instead, the engineer will use bid abstracts (or bid tabulations) to help him prepare this estimate.

Bid abstracts show contract prices from other jobs that have been awarded by government agencies. The designing engineer will try to find the cost of completing a project similar to the one he's planning. When he does, he'll adjust the cost per linear foot up or down as seems appropriate and then multiply by the total length of the current project. This isn't the best way to estimate, but it does establish a projected cost. *You'll* have to do much better.

Bid abstracts are available to engineers, contractors and the general public for nearly all work that's publicly funded. Figure 1-1 shows a bid abstract.

Bid abstracts show unit prices — usually the cost per linear foot of utility line installed. In theory, if you were to average all the bids submitted for each item of work, you'd have a good average price for that item. But it doesn't work that way. Many bids are unbalanced, being padded (too high) for some items and showing unrealistically low prices for others. This distorts the picture. You'll find more discussion of unbalanced bids (and an example) later in this chapter.

There's a reason for unbalanced bids, of course. No contractor is anxious to have his competitors know his real costs. Also, padding the price of work that will be done early in the project can fatten progress payments that are to be received early in construction. You're not supposed to do that, of course, but it happens because contractors like the comfort of having extra cash to work with. I can't blame them.

But bid tabulations can still be a useful tool. Use them to cross-check prices on items that are unfamiliar to you. If the bid tabs you use aren't for jobs awarded within the last few months, allow for changes in labor and material prices since the estimate was prepared.

Tabulation of Bids Submitted for City Sewer Project 275
Date: July 1, 1987   Time: 3:45PM   Place: City Hall

Item	Approx. Qty.	Unit	Firm Engineer Unit Price	Ext.	Firm Jones Const. Unit Price	Ext.	Firm Smith Const. Unit Price	Ext.	Firm Green Const. Unit Price	Ext.	Firm Nolan Const. Unit Price	Ext.
310.03.1 Trench Excavation Type I & Type II Backfill (Type I)	1,090	LF	23.00	25,070.00	11.00	11,990.00	19.50	21,255.00	12.00	13,080.00	25.00	27,250.00
310.04.1 Type I Bedding Material (1½" Drain rock)	1,090	LF	2.00	2,180.00	2.00	2,180.00	5.00	5,450.00	4.00	4,360.00	3.00	3,270.00
311.04.1 Type "P" Surface Restoration	1,120	LF	15.50	17,360.00	15.00	16,800.00	8.00	8,960.00	22.00	24,640.00	23.00	25,760.00
505.02.1 Sanitary Sewer Pipe Size 8"	1,090	LF	7.50	8,175.00	8.00	8,720.00	2.50	2,725.00	10.00	10,900.00	9.00	9,810.00
505.02.1 Sanitary Sewer MH's	4	ea	1,000.00	4,000.00	1,100.00	4,400.00	1,400.00	5,600.00	1,500.00	6,000.00	1,100.00	4,400.00
505.03.1 Service Lines Size 4"	675	LF	12.00	8,100.00	12.00	8,100.00	15.00	10,125.00	15.00	10,125.00	7.00	4,725.00
505.03.1 Service Lines Size 6"	50	LF	15.00	750.00	14.00	700.00	9.00	450.00	50.00	2,500.00	9.00	450.00
505.03.2 Service Line Connections (8" x 4")	27	ea	100.00	2,700.00	100.00	2,700.00	75.00	2,025.00	125.00	3,375.00	50.00	1,350.00
505.03.2 Service Line Connections (8" x 6")	2	ea	120.00	240.00	100.00	200.00	95.00	190.00	175.00	350.00	60.00	120.00
505.04.1 Standard 8" Temporary Cleanout	1	ea	250.00	250.00	300.00	300.00	250.00	250.00	1,000.00	1,000.00	350.00	350.00
TOTALS				68,825.00		56,090.00		57,030.00		76,330.00		77,485.00

Bid abstract
Figure 1-1

Several weekly and monthly magazines show bid tabs and current price trends. Figure 1-2 shows a comparative price list and bid tab taken from *Pacific Builder & Engineer*. The comparative unit prices give you an idea of the range of prices.

Several estimating guides are available to help you estimate prices. These guides show material, labor and equipment costs, and typical manhours for most work. But be careful when using estimating guides. The cost of utility line work can vary drastically. Installing utilities in a busy city street requires different tools, equipment, labor and materials than you'll use in a rural area. That makes the cost very different.

Estimating guides are better suited to other trades in the construction industry, such as structural work. If there's a pump station on your project, an estimating guide may be useful. But costs for the underground portion of the job depend too much on variables unique to every job. The best guide to costs on your next job will always be costs on the last several projects you've completed — plus your judgment, of course. Use labor, material and equipment costs and labor production rates from your own cost records whenever possible.

Working with the Engineer
The contracting authority selects the engineer. It's the engineer's job to make sure you do the work right. He'll be responsible for the work you do long after your warranty period ends. This means he'll want to monitor your work very carefully. This is no problem, provided he knows his business. He should be tough but fair. An engineer who isn't a competent professional may try to make you the scapegoat for his mistakes. Avoid working with engineers or engineering firms that have a reputation for ruining contractors and construction projects. If you end up in court, the only winners will be the attorneys.

Before you decide to bid a project, check into the reputation of the engineering firm. Ask other contractors what they know about the firm. Do they have a history of lawsuits, contractor failures and incomplete projects? How much experience do they have? If you have any doubt about the competency or professionalism of the designing engineer, beware. Protect yourself. Either don't bid the work, or else cover yourself with a high margin.

WATER SYSTEM
Poulsbo, Washington

The Poulsbo City Council opened bids on March 25 for construction of 6,134 ft. of barbed wire and chain link fence, pumphouse modifications, construction of 1,290 ft. of 8 in. water line and 200 ft. of asphalt pavement for Big Valley well and water improvements. The three lowest bidders and their unit prices for principal quantities follow:

1. Ryan Built Inc., Port Orchard, WA, $126,858.18.
2. Blue Cascade Construction Inc., Port Orchard, WA, $132,134.10.
3. Blossom Construction, Bainbridge Island, WA, $133,951,91.

1,048 LF Furn & Inst Chain Link Fencing w/3-Strand Barbed Wire	7.60	8.00	7.95
1,102 LF Furn & Inst 4-Strand Barbed Wire (Stl Posts)	1.62	2.25	1.70
2,551 LF Furn & Inst 3-Strand Barbed Wire (Stl Posts)	1.40	2.00	1.50
1,433 LF Furn & Inst 4-Strand Barbed Wire (Wood Posts)	2.64	3.00	2.75
LS Well Site No. 1 Sub Well Pump	9,425.00	8,500.00	8,500.00
LS Well Site No. 1 Wellhead Protect Structure	1,826.00	3,500.00	4,200.00
3 EA Well Site No. 1 Sump & Manhole Locking Covers	318.00	325.00	500.00
LS Well Site No. 1 Pumphouse Modif	6,000.00	7,000.00	4,500.00
LS Collection Sump Overflow Line, etc	720.00	1,550.00	1,200.00
LS Well Site No. 1 Pumphouse Modif (Pump Rplcmnt, relocate; piping)	12,395.00	12,500.00	10,500.00
LS Little Valley Booster Sta Modif	8,505.00	8,500.00	9,800.00
LS Furn & Inst Pump Cntrls, Elec Serv	18,255.00	18,000.00	18,400.00
LS Relocate & Extend Elec & Telephone	1,034.00	1,500.00	1,500.00
LS Well Site No. 2 Grading	450.00	450.00	400.00
1265 LF Furn & Inst 8" Cl 50 Cem Lined DI Wtr Pipe	12.75	12.50	15.00
4 EA Furn & Inst 8" Gate Vlv	372.00	450.00	500.00
1 EA Furn & Inst 8" Check Vlv	462.00	550.00	1,100.00
1 EA Furn & Inst Air Relief Vlv Assy	279.00	450.00	380.00
1 EA Furn & Inst Blow Off Vlv Assy	325.00	650.00	450.00
102 CY Rdwy Excav, Embank Constr	20.00	5.00	8.00
230 TON Furn & Place Base Course	6.50	6.00	12.00
80 TON Furn & Place Top Course	12.60	10.00	14.00
35 TON Furn & Inst Asph Concr Pvmt	104.00	60.00	40.00
LS Erosion Control	1,000.00	1,000.00	350.00
LS Furn & Inst Cattle Guard	5,000.00	2,500.00	5,100.00
2 EA Reinforced Concr Headwall	250.00	725.00	600.00
LS Concr Sidewalk & Grvl Path	442.00	750.00	700.00
LS Gravel Access Rd Constr	3,800.00	1,600.00	1,600.00
LS Spring Box Constr & Renovation	4,900.00	9,500.00	9,000.00
25 LF Furn & Inst 8" Cl 150 DI Wtr Pipe	27.00	40.00	40.00
1 EA Furn & Inst 8" Gate Vlv	372.00	500.00	700.00
1 E Furn & Inst Blow Off Vlv Assy	559.00	750.00	450.00

Courtesy: Pacific Builder & Engineer

Trade magazine price list and bid tab
Figure 1-2

SEWER
Missoula, Montana

The Missoula County Clerk opened bids on July 13 for construction of 6,860 ft. of 12-in.–15-in. PVC sewer and 2,690 ft. of PVC force main in Reserve St. and Grant Creek to serve Sunlight Properties. The three lowest bidders and their unit prices for principal quantities follow:

1. L.S. Jensen and Sons, Inc. $333,637.65.
2. American Excavating, Inc. $361,402.85.
3. R.H. Grover, Inc. $466,543.00.

1,880 LF 12" SDR 35 PVC Swr Line	16.35	16.10	26.70
4,980 LF 15" SDR 35 PVC Swr Line	17.64	18.35	24.20
140 LF 6" CI 160 PVC Force Main	9.74	10.00	16.00
2,550 LF 10" CI 160 PVC Force Main	9.38	12.20	14.50
85 LF 6" CI 50 DIP	16.78	10.00	20.00
55 LF 12" CI 50 DIP	17.98	26.35	22.80
185 LF 24" CI 150 DIP	45.41	60.20	73.00
85 LF 10" CI 50 DIP	29.38	12.00	38.60
25 EA Stndrd 48" Dia Mnhl	692.00	685.00	710.00
240 LF Ext Depth Stndrd 48" Dia Mnhl	52.00	70.00	53.00
2 EA Stndrd 72" Dia Mnhl	2,215.00	2,955.00	1,160.00
26 LF Ext Depth Stndrd 72" Dia Mnhl	100.00	40.00	113.00
1 EA Stndrd 72" Dia Drop Mnhl	6,318.00	6,965.00	5,045.00
11 LF Ext Depth 72" Dia Drop Mnhl	100.00	40.00	151.00
LS Bored Casing at Mullan Rd	34,775.00	38,710.00	35,000.00
3 EA Force Main Cleanout Compl	1,100.00	1,475.00	1,226.00
LS Pkg Lift Sta No. 1 Compl	44,153.00	43,590.00	81,484.00
LS Pkg Lift Sta No. 2 Compl	35,729.00	43,120.00	53,533.00
1,825 CY Embankment	1.60	.50	2.28
34 LF 36" Dia RCP Wall B CI 2	40.00	46.65	52.00
6 LF 12" Dia 16 Ga. CMP	13.75	38.00	25.00
1 EA Force Main Emrgncy Pump Conn	2,162.00	1,506.00	2,243.00
600 CY Cushion Base	5.00	10.30	7.00
LS Casing Grouting	2,140.00	930.00	2,300.00
23,700 SF Stabilization Fabric	.11	.14	.25

Courtesy: Pacific Builder & Engineer

Trade magazine price list and bid tab (continued)
Figure 1-2

When you've signed the contract, work must be completed at any cost. Walking off a job usually means you're out of business. Remember, every bid is, in essence, a firm commitment to enter a contract. If you withdraw your bid or fail to enter into the contract, the owner will demand that you pay a stiff penalty or request that your bid bond be forfeited. That's why you should select your jobs carefully.

Knowing Your Costs

As an independent contractor, you have the freedom to charge any price you want. The catch is that if your price isn't competitive, you'll never have any work. To win the contract and make a profit, you have to bid accurately. That's possible only if you know your costs. Guessing will get you nothing but trouble.

Your costs include equipment, labor, materials, subcontracts, and taxes, of course. These are sometimes called *direct costs* because they're the direct result of taking some specific job. If you hadn't taken that job, you wouldn't have these costs. But just as important and just as real are the indirect costs. These are also known as *overhead*: office staff, rent, heat, light, advertising, insurance, auto and truck expense, and all the other costs that every business has, no matter what their product. These costs go on even if you didn't have the specific job in question. But don't think they aren't real. They're very real indeed and have to be included in every job you have.

Last, and most important, comes your markup. That includes your profit, the reason for doing the job. Don't forget it!

Let's take a detailed look at each of these costs and how they come together in your estimates. Then we'll look at the direct cost sheets you'll use to write up your estimate.

Tools and equipment— If you can afford to buy all the tools and equipment you need for installing utility lines, you probably don't need to work. Owning several million dollars worth of heavy equipment would make you independently wealthy. Sell the stuff, invest the proceeds and retire on the interest.

Utility line contracting is capital intensive. It takes money to make money, unlike homebuilding or remodeling where you can load all the tools and equipment you need into the back of a truck at the end of the day. In this business most jobs will require equipment that costs hundreds of thousands of dollars. No one is going to start out owning that much equipment. In fact, even many of the big boys, contractors with million dollar projects and dozens of employees, don't buy their equipment. They can't afford to. They rent or lease most of what they need, buying only the less expensive equipment that's sure to be in use nearly all the time. This trend seems likely to continue. Most equipment dealers offer lease-purchase plans. In any case, you can nearly always rent what you need for the time you need it. Buying equipment should be the last thing you consider until you're handling a heavy volume of work nearly all the time. Any equipment you buy reduces your working capital.

The daily costs of short-term rentals will be relatively high because the dealer has higher overhead and maintenance costs for short-term customers. Also, there's a better chance that the equipment won't be rented out every day if it's offered for short-term rental.

My advice for the beginner is to ignore the higher daily costs for day-by-day rentals. Whenever you can, rent equipment for the exact number of days that you need it. When construction slows down, you won't be stuck making payments on expensive equipment that you're not using.

Whether you rent, lease or buy your tools and equipment, you must figure out the *daily operating costs.* That's just part of knowing your true costs. Entire books have been written on how to calculate equipment operating costs. You don't need to know that much to be successful in this business. Here's all you need to know about equipment operating costs.

If you rent or lease your equipment, it's fairly simple to calculate your daily operating cost. Add together your rental fee (either daily, weekly or monthly), fuel and service costs. Divide this total by the number of working days. The answer is your *daily* operating cost. To get an *hourly* operating cost, just divide the daily cost by the number of working hours in a day.

Owning and operating costs— If you own the equipment, figuring your operating costs is a little harder. Your costs include depreciation, taxes, insurance, repairs, fuel, and servicing costs. Figure 1-3 shows some typical hourly operating costs for the kinds of equipment you'll be using. These rates are derived from monthly rental rates, and may vary from area to area.

As a general rule, *owning* and operating costs for this equipment will be slightly less. But remember, ownership costs will vary dramatically depending on how you use the equipment. For example, your hourly O & O costs will be lowest when you can use the equipment for a full construction year of 1,600 hours. The less you actually use the equipment, the higher your hourly costs will be.

But this is still an optimistic owning and operating cost estimate. It doesn't include any interest you may be paying on the initial purchase price of the equipment. And it assumes you

Machine	Size	Hourly operating cost
Wheeled backhoe	50- 70 hp	$12-15
Wheeled backhoe	70-100 hp	$18-20
Tracked excavator	40,000 lbs, 120 hp	$30-35
Tracked excavator	60,000 lbs, 160 hp	$40-45
Wheeled loader	160 hp	$28-32
Wheeled loader	250 hp	$45-50
Dozer	70 hp	$12-15
Dozer	150 hp	$35-40
Dozer	200 hp	$55-60
Sheepsfoot roller	5' drum width	$30-35
Trash pump	4", 600 GPM	$ 5-10
Trash pump	6", 1000 GPM	$10-15

Typical hourly operating costs
Figure 1-3

won't have any major accidents or unusual breakdowns during the five years of use. Note that there's no profit for you in the costs in Figure 1-3, and that they don't include the operator's wage. Profit and labor will be added later.

Damage and hard usage: Repairing major damage on large equipment can cost you a lot of money. Machines roll over. Materials fall on top of them. A careless operator can puncture your brand new $500 tire. Vandals can sabotage your equipment. You can get insurance to cover these kinds of damage. I recommend it. Working conditions will also affect the amount you spend on major repairs. If you're working in soft soil and there isn't much dust, your backhoe will have a longer and less costly working life. If you're working in hard soil and in dry, dusty conditions, your operating costs will be 8% to 10% higher. If your basic operating cost is $24.70 per hour, your hard usage operating cost may be 10% higher, $27.17 per hour.

Today's equipment is designed and built for hard usage. If you maintain your equipment properly, it can probably be run at maximum capacity under hard usage conditions without doing serious damage. I'll run my equipment hard to complete a job on schedule or to cut my labor cost. Other contractors baby their equipment, even if it slows down production. To my way of thinking, that isn't usually necessary. But if it's your equipment, you be the judge.

Now you've calculated your hourly equipment cost. But how do you figure the amount of work a machine can do in an hour? That's the most important consideration when pricing any job. Unfortunately, estimating equipment productivity isn't easy. That's why I'll devote all of Chapter 2 to that topic. I'll show you the most accurate way I know to estimate equipment performance.

Estimating Labor Costs
Your next major cost is labor. Labor is your most controllable job cost. Your hourly labor cost will usually be set by a combination of labor agreement and government regulation. Be sure to check your contract for wage and benefit requirements. And remember that the hourly labor rate is only part of the picture. Payroll taxes and insurance, which I'll cover shortly, will add another 20 to 30 cents for every dollar of payroll.

If the wage rate is set by contract, your hourly labor cost is a fixed cost. But a contractor who hires the right tradesmen and provides good supervision will end up spending less on labor than a contractor who hires slow or inexperienced tradesmen. Minimize your labor cost *per unit* by using motivated, skilled tradesmen. Remember, it's your most controllable cost.

Chapter 2 explains how to estimate labor and equipment productivity per manhour. It's always hard to predict labor costs accurately — weather, illness, working conditions and routine construction delays will affect your labor costs on most jobs. But the more you know about your crews and the more accurate your records of their past performance, the more accurate your labor estimates are likely to be. I'll cover cost record keeping later in this chapter.

There are good production rate estimates and there are bad production rate estimates. That's unavoidable. But there are some estimates that simply cannot be. Those are very avoidable. It should be easy to see what's simply *not possible*. If a machine can move 110 cubic yards of dirt in an hour, don't figure you're going to move 880 cubic yards in an 8-hour day. You'll never put 8 hours of 100% pure production back to back. That just doesn't happen.

It's always good practice to visit the job site before drawing up your labor estimate. All your calculations may suggest an installation rate of 300 feet per day. But when you visit the site, you may discover problems that will make even 150 feet per day an optimistic estimate. Overhead lines, large trees, restricted work room, unavoidable damage to property — all will lower your production rate well below projections. I'll cover these and other similar variables in detail in the next chapter.

Your Labor Burden

As I just said, your hourly labor cost is far more than what you pay employees. The contractor's "labor burden" will add between 20% and 30% to all of your labor costs. For every dollar of payroll, you must pay an additional 20 to 30 cents in taxes and insurance to government agencies and insurance carriers. That 20% to 30% will usually be more than your profit in the job. Here's a breakdown of where all that money goes.

Unemployment Insurance— Most states levy an unemployment insurance tax on employers, based on the total payroll for each calendar quarter. The actual tax percentage is usually based on the employer's history of unemployment claims, and may vary from less than 1% of payroll to 4% or more.

F.U.T.A.— The Federal government also levies an unemployment insurance tax based on payroll. The tax has been about 0.8% of payroll.

F.I.C.A.— The Federal government also collects Social Security (F.I.C.A.) taxes. This comes to about 7.15% of payroll, depending on the earnings of each employee, and is collected from the employer each calendar quarter, or less.

Workers' Comp— States generally require employers to maintain workers' compensation insurance to cover their employees in the event of a job-related injury. Heavy penalties are imposed on employers who fail to provide the required coverage. The cost of this insurance is a percentage of payroll and varies with the type of work each employee does. Clerical and office workers have a very low rate. Your cost may be less than 1% of payroll. Hazardous occupations such as roofing — and sewer installation — carry a rate closer to 25% of payroll. Equipment operators and truck drivers usually have a rate between 6% and 12% of payroll.

The actual cost of workers' compensation insurance varies from state to state and from one year to the next, depending on the history of injuries for the previous period. Your insurance carrier can quote the cost of coverage for the type of work your employees are doing.

Liability Insurance— Every contractor should maintain liability insurance to protect his business in the event of an accident. There are two types of liability coverage: personal injury and property damage. Both are based on your total payroll. A comprehensive liability package with coverage up to $200,000 per accident will be about 5% of payroll. Higher liability limits cost more.

The total contractor burden can be itemized as follows. The percentages listed are approximate maximums. Your accountant or bookkeeper will have more exact figures.

State unemployment insurance   4.0%

F.I.C.A. and Medicare   7.2%

F.U.T.A.   0.8%

Workers' compensation insurance   6.0% to 12.0%

Liability insurance   5.0%

Total contractor burden   23.0% to 29.0%

Can you skip paying some of these taxes? Not hardly. There's no legal way to avoid paying 100% of the taxes outlined above. If you have payroll, you have to add this labor burden into every estimate and make the insurance and tax deposits when due. No contractor can ignore these requirements and operate for long.

How do you add these costs? Easy. If your labor burden is 25% of payroll, add 25% to the total hourly cost (including fringe benefits). For example, if your tractor operator makes $20 per hour including base pay and benefits, your hourly cost will be $25 (1.25 times $20). Taxes and insurance will cost you, the contractor, $5 per hour. You won't have to pay taxes and insurance on certain fringe benefits. Your accountant can supply more exact figures.

Some contractors add their labor burden to the estimate after all other costs are tabulated. There's nothing wrong with that system — if it works for you and if you don't forget this important item. I prefer to include the labor burden in each hourly or daily labor cost. That way I never forget it. When you see any labor cost in this book, you can assume that it includes the base rate, fringe benefits, and all taxes and insurance. But beware. Wage rates and employment costs vary widely from area to area.

Estimating Material Costs
Publicly advertised projects will attract the attention of your local material suppliers. Major suppliers have employees who will do the material take-off for you at no charge. The supplier makes up a materials list for the job and quotes a price for all materials he can provide. His hope, of course, is to sell the materials if you're awarded the job.

Many of these take-off people are very professional and can be expected to produce accurate material lists. But don't put blind faith in their work. It's too easy to accidentally leave out one section of the project, or to make an error in addition. Even experienced professionals make mistakes. And the more complex the job, the more room there is for error.

More than one company will probably quote on the materials needed for each contract. Cross-check the take-off sheets from different suppliers. Compare their quotes. Make sure nothing has been left out. But it's not ethical to tell one supplier what his competitor's prices are. If you call a supplier, reveal his competitor's quote and ask if the price can be trimmed, it probably will be. But the small advantage you'll gain on this one bid isn't worth the aggravation on future bids. Bid shopping makes it harder to find suppliers willing to quote prices before the day of bid opening.

Be wary of incomplete quotations or bargain prices on one or two items. It's tempting to split your order among several suppliers to take advantage of isolated bargain prices. But splitting the order increases your overhead and doesn't give you as much leverage with any one supplier. In the long run that probably means higher costs for you.

Some suppliers will want a commitment from you in return for their take-off and price quotation. They want as complete an order as possible — even before you get the contract — for everything they can supply for the job. But my advice is to avoid making any commitments before you have the job. Most suppliers are more willing to negotiate when you're ready to place a major order.

If the project you're bidding isn't a large job, don't be surprised if no material supplier is willing to do the take-off. You'll have to prepare your own take-off and gather quotes on materials.

Collecting Subcontract Bids

Every prudent utility line contractor is cautious when dealing with material suppliers. And it's equally important to be cautious with subcontractors. An unreliable subcontractor can get you into deep trouble in a hurry. A sub who can't do what's promised at the price quoted will jeopardize your whole contract. Part of the foundation of the bid will have washed away, leaving you with a serious problem. And if he does a poor job, it's *your* reputation that suffers. The general contractor is always responsible for the work of his subs. Solicit subcontract bids from trustworthy and competent subcontractors.

But even with the best subs, double check each bid for accuracy before using it in your bid. Look out for sub prices significantly below the competitors.'

We've looked at the equipment, labor, material, and subcontract costs that make up your direct operating cost. Now let's look at the last items you'll add to most bids: overhead, contingency, escalation and profit. Together, these items are usually called *markup*.

Overhead

We've looked at four of the most important costs for a utility line contractor: equipment, labor, material and subcontracted items. Just as important, but perhaps less obvious, is overhead. Start with the understanding that every cost is important when preparing an estimate. You should be just as anxious to identify and price overhead costs as you are to price labor and material costs. Remember, good estimators identify every cost and put a price down beside every cost they identify. Your estimate for anything missed is always zero. That's a 100% miss every time.

I'll divide overhead into two categories, direct overhead and indirect overhead.

Direct Overhead
On every job you have costs that aren't associated with any particular trade or phase of construction, but are the result of taking that particular job. These costs are usually called *direct overhead* and can be thought of as administrative costs. They aren't labor, material, equipment or subcontract items. In fact, many direct overhead items don't show up in the plans or specs. You have to find them and price them yourself.

The list below includes the items that are usually included as direct overhead. It's a pretty complete list, and can save you hundreds of dollars on the next few jobs. My advice is to review this list before you complete every estimate. Naturally, not every job has every cost item listed below. But on most jobs, reading over this checklist will help you discover several items you had forgotten.

Bonds (bid, completion, maintenance, street encroachment, street repair)
Debris removal (trucking, dump fees)
Dust protection
Expendable tools
Field office (storage, tool crib, watchman's)
Field supplies
Insurance (workers' comp, property damage, bodily injury, fire, builder's risk, equipment floater)
Licenses (business license, state contractor's license)
Mobilization and Demobilization
Moving utility lines
Permits (blasting, building, sidewalk, street obstruction, Sunday work, temporary, wrecking, debris burning)
Photographs
Protection of adjacent property
Repairing damage
Signs
Site inspection
Supervision (superintendent, foreman, engineer, timekeeper, payroll clerk, material checker, watchman, supervising subs)
Surveys
Taxes (excise, payroll, sales)

Telephone
Temporary fences
Temporary utilities (power, water, job toilet)
Traffic control (barricades, flaggers, lane delineators)
Travel expense
Weather protection

You can probably think of other direct overhead items. Most contractors include the cost of supervision and other nonproductive labor, such as the cost of estimating the job. In my opinion, the time you spend on each job should be charged against each job. Other contractors include all office work under indirect overhead, even if the work relates to some particular job. I won't say that's wrong. It's just not the way I do it. The only important thing is to remember that these are very real costs and must be included somewhere in every estimate. Since they're the result of taking a particular job, they fit nicely under direct overhead for that job.

Indirect Overhead
Even after all job costs are compiled, there's still more overhead to include. Every business has expenses that can't be charged directly against any particular job. Here are some examples:

Advertising
Car and Truck
Depreciation
Donations
Employee medical benefits
Licenses and Fees
Maintenance and Repair
Office insurance (fire, liability, workers' comp)
Office rent
Office telephone
Office utilities (water, power, gas, sewer)
Office staff (clerical, management)
Payroll taxes
Pension and Profit sharing
Postage
Professional fees (accounting, legal)
Stationery and Supplies
Travel and Entertainment
Uncollectibles

These are all *indirect overhead* or office costs. They differ from direct overhead because they go on even when work in the field stops.

In the utility line contracting business it's important to keep your overhead low — no more than 10% of gross receipts. If your overhead is too high, you won't be competitive.

Here's where being a small contractor gives you an edge over the larger contracting companies. They have fancy offices staffed with full-time personnel. It's just about impossible for them to keep their overhead low. If you can run a compact, efficient, low-overhead operation, you have the advantage at every bid opening.

Including Overhead in Your Bid
It's fairly easy to include direct overhead in your bid. Just figure the cost of supervision, insurance, mobilization, etc. and add those costs into your estimate. Indirect overhead is harder. In fact, there's no single correct way to figure your indirect overhead cost on any job. Having admitted that, I'll suggest the system *I* use.

I start by estimating my overhead for the entire year. That's not too hard. I've been involved in the business for many years and know about how much I'll spend on office rent, heat, telephone, etc. Then I divide the estimated indirect overhead by my estimated gross volume for the year. Again, I can usually make a pretty good guess what my billings will be for the year. The answer is my indirect overhead percentage for all jobs I estimate for that year.

Here's an example: If my gross is expected to be $1,000,000 and my indirect overhead will be about $100,000 (including my salary for running the business), my indirect overhead should be 10% of every bid.

Other contractors divide annual indirect overhead by 40 or 50, depending on the number of productive weeks each year, and add a week's overhead to every job for each week it's expected to last. In my example, $100,000 divided by 50 is $2,000. If you have one job and it's expected to last one week, add $2,000 for indirect overhead. If you're doing two jobs of about the same size during a given week, each would bear one-half the indirect overhead cost for that week. Multiply by the duration of the project and you have the indirect overhead cost for that project.

Some contractors figure the indirect overhead cost as a cost per productive manhour. Other contractors have reduced the indirect overhead to a cost per linear foot of line laid. Any of these systems is good if it works. Do what makes the most sense to you. But the important thing is to keep a record of your indirect overhead and develop some method of dividing this amount among your jobs.

Don't forget to recover your job overhead and office overhead *on every job and in every estimate.* The total cost of direct and indirect overhead will usually be about 10% of the total cost on nearly every job. Forgetting overhead in your bid will more than wipe out the profit on just about every job you take.

Before I leave the subject of overhead, let me make one more important point. If your overhead is 10% of gross receipts, you can't simply add 10% to each bid to cover overhead. Look at this example.

Suppose I'm in the apple business, selling apples from a sidewalk stand. My overhead runs 10% of gross sales. How much do I have to mark up the apples to cover overhead? If you guessed 10%, you're wrong. Here's why:

Say my selling price less overhead is $90. If I add 10% to cover overhead, I'll sell those apples for $99. That's not enough. You can see that I have to sell them for $100 to recover all of my 10% overhead. By adding 10%, I'm recovering $1 less than my actual overhead cost. What's the answer? Easy. Instead of multiplying by 10%, divide by 90%. Divide $90 by 0.90 and you get $100 exactly. Overhead is 10% and my selling price recovers the full $10.

Use the same system with your estimates. If overhead is 11%, divide by 0.89 to find the cost plus 11% overhead. Strangely enough, it doesn't make any difference whether you do this before or after factoring in other markup. The result will be that overhead is 11% of the bid price.

Vary the Markup with the Job

Now that I've made figuring markup so easy, I'm going to complicate it a little. Let's use a simple example of a $100,000 job — that is, one whose total direct job costs amount to $100,000. We've decided to add 10% for direct and indirect overhead and 10% as a target profit. We bid the job at $120,000.

It's not really quite that simple in real life, however. The concept of a flat 10% doesn't allow for covering risks or for rating the varying proportions of the bid contents. There are other methods of adding markup.

For a moment let's imagine that you're a large general contractor engaged in contracts worth millions of dollars a year. You bid a $2 million building contract, and intend to follow the current trend to use subcontractors for almost all the work. If the total subcontract and overhead amount to $1.8 million, you would have a target profit of $200,000.

But look at that bid again. Since your own costs are fairly stable, you may be willing to cut the profit margin a little to get the contract. In theory at least, not a a whole lot can go wrong. You may decide to add just 8%. In fact many large general contractors operate on a margin smaller than 8%. When costs are stable and the general contractor runs an efficient operation, a small percentage of profit off a large volume of business amounts to a hefty chunk of change.

Now go back to our $120,000 bid for a minute. We'll look at two scenarios for that job. In the first, the materials for the job will cost $60,000. An asphalt contractor wants $10,000.00 to repair the street. That totals $70,000. The job will take about two weeks. Our portion of the work is about a $30,000 job, and we've marked it up $20,000. In the second scenario, it's a 10-week job, and our labor portion amounts to $70,000. We'd have a $20,000 markup on $70,000 worth of what may be high risk work.

To try to compensate for such differences in job makeup and risk, estimators came up with the 20-10-10-10 or the 15-15-10-10 method of adding markup. Use a higher markup on the first two cost categories (labor and equipment) and a lower markup on the other two cost categories (material and subcontracts). The part that has the higher risk, your labor and equipment, should carry more markup. It's a lot chancier than the fairly stable fixed cost of materials and subcontract items.

Let's see how this system would affect the bid in the two jobs I've described.

Job 1

Labor and equipment: $30,000 plus 20% = $36,000
Materials and subcontracts: $70,000 plus 10% = $77,000
  Total bid: $113,000

Job 2
Labor and equipment: $70,000 plus 20% = $84,000
Material and subcontracts: $30,000 plus 10% = $33,000
  Total bid: $117,000

There are two important points to grasp here. First, your overheads are real and must be included. Second, they're not a constant ratio of each and every bid.

It's worth the time and effort to look at your overheads and try to place them where the costs are incurred. If you bid the $100,000 job at $120,000 and a competitor bid at $117,000 because he understood his operating costs better than you understand yours, he'd win the job by less than 3%. Remember, in competitive bidding there's no second place winner.

As a general rule, if you bid at the total of labor, equipment, material, and subcontracts plus 10%, you're barely covering your direct job cost overheads. Unless your overheads are minimal, your profitability is at risk.

I can't begin to suggest a rule of thumb, but here's some general advice. If anything about a job bothers you — you don't like the engineer or the owner, you have some doubts about the general contractor, you consider the risks very high, you plain don't feel comfortable with your projections, your unit prices look low in comparison to bid tabulations on similar jobs, you already have a good work load — *mark it up*. Mark it up 12%, 15%, 20%, or 35%. When in doubt, select a higher number. Feel comfortable with your bid going in. You may get that job. Once the contract is signed, you're on the roller coaster and nobody will let you off until the ride is over.

Contingency and Escalation
Most contractors add a small amount to their bids to allow for the unexpected. That's good practice on utility line work and all excavation. Most surprises on a utility line jobs will increase costs, not decrease them. Your allowance for *contingency* leaves a cushion to fall back on.

The right amount to add for contingency depends on the contractor and the job. Many excavation contractors routinely add 2% or 5% to their bids. Pipe jacking and tunneling might require larger allowances to meet difficulties which can't be forecast accurately before work begins.

A word of caution here. Contingency isn't intended to cover for sloppy estimating. True, it's common to have a large contingency allowance in preliminary estimates that are made before the final bid is prepared. But you won't be bidding jobs like that. If you can't figure out what the job requires, either get more information or don't bid. Don't use a contingency allowance to cover for what you don't understand.

Escalation is the increase in costs of labor, materials and equipment between the time the bid is submitted and the time work is actually done and paid for. Even though you're sure of the cost of labor when you submit your bid, you may not know what you'll be paying equipment operators when they do the work. This is very important during high inflation periods.

If the job is expected to continue for several months, and if there's a good chance that labor, material or equipment costs will increase before the job is completed, consider including a small allowance for escalation. If you can't get firm quotes for materials to be delivered in the future, either allow for price increases or specifically exclude price increases from your bid. Qualified bids are usually not allowed in public utility work. But you can often qualify a subcontract bid to a general contractor.

The magazine *Engineering News Record,* published by the McGraw-Hill Book Company at 1221 Avenue of the Americas, New York, N.Y. 10020, follows price trends in construction materials, labor and equipment, and can help you follow price changes. *Construction Review,* published by the United States Government Printing Office, Washington D.C. 20402, follows prices of major groups of construction materials.

Watching Your Profit
The profit is the return on the money you have invested in your business. It isn't your pay for doing the work you do. You should get a wage for the work you do and, in addition, receive a return on the money invested in your business.

If a contractor has $50,000 invested in his business, he should receive a return on investment of $3,200 to $6,000 per year (8% to 12% of investment) in addition to a reasonable wage. This profit can be thought of as interest on the money invested in equipment, office, inventory, work in progress and everything else needed to run a utility line contracting business.

How much, then, should you include in your estimate for profit? You'll hear many conflicting figures. Some estimators say a 20% profit is a good target and they try to end up with a "profit" of 20% of the total contract price after all bills are paid. Some utility line contractors may operate efficiently enough to earn a 20% profit. But they certainly are the exception. The contractor who talks about a 20% profit may mean that he has 20% left after paying for labor, materials and equipment. Most of this 20% is probably needed to cover overhead and the contractor's wage. That's not profit in the true sense. A profit is what remains after *all* costs are considered. The cost of your own work should be included in your estimates either under direct or indirect overhead (or maybe both).

What then is a realistic profit in the true sense? Dun and Bradstreet, the national credit reporting organization, has compiled figures on construction contractors for many years. They report the average net profit after taxes for all contractors sampled to be consistently between 1.2% and 1.5% of gross receipts. This includes many contractors who reported losses or became insolvent.

A 1½% profit, even after taxes, is a fairly slim profit. Not many contractors include so small a profit in their bid. On extremely large projects such as highways, power plants or dams, a contractor may include only 1% or less for profit — especially on a "cost plus" contract where there's little or no risk of losing money on the job. Residential construction, especially remodeling and repair work, traditionally carries a higher profit margin because the size of jobs is much smaller and the risk of loss is larger.

Of course, there's more to profit than just how much profit you would like to make. Sharp competition will reduce the amount of profit you can figure into your estimate. If you include too much profit in your bids, you'll find yourself underbid for the jobs you'd like to have. If you've developed a specialty, do a particular type of work better than other contractors, and have enough work to stay busy, then it's appropriate to increase your profit by a few percent.

In practice, there's no single profit figure to fit all situations. For most utility line work, an 8% to 10% profit is a very nice expectation. But even a 5% profit may be enough if your bid covers all costs, including direct and indirect overhead. And when work is plentiful and the job suits your experience and capabilities nicely, a 15% profit may not be excessive.

High-risk jobs should carry higher profit margins. If you have
the know-how and equipment to handle the difficult jobs, you
can normally demand and get a premium. But be aware that the
difficult jobs usually come with complicated, unexpected,
expensive problems. Even an estimated profit margin of 20% or
30% can evaporate pretty quickly when problems begin to delay
the work. Keep this in mind when you're deciding whether or
not you want to bid high-risk jobs.

Every contract has risks, but some projects have risks that are
guaranteed to cost you money. Make sure you cover these risks
in your estimates. If you're working in a hilly, boggy area,
there's a good chance your equipment will get stuck or
overturned. If the project requires deep excavation in an area
with groundwater, expect that shoring and dewatering problems
will slow production and require special equipment. If the job
requires blasting, sections of unfractured rock may have to be
reshot. Keep that in mind when you're figuring contingency and
profit.

Here's the acid test of your profit percentage. Are you
earning a reasonable return on the money you have invested in
the business (after taking a wage for yourself and paying all
overhead expenses)? If you are, then your profit estimates are
probably about right. If not, some adjustment is necessary.
Each year you should earn a profit equal to 8% to 12% of the
"tangible net worth" of your business. Tangible net worth is
the value of all the assets of your business less the liabilities
(anything your business owes) and less any intangible items such
as goodwill.

Where the Markup Goes
Overhead, profit, escalation and contingency aren't usually
itemized in your bid. They don't appear as separate items in the
bid you submit to the contracting authority. Instead, these costs
are distributed among all bid items by marking each up by some
percentage. Here's how to do the calculation.

Add together your direct costs and markup (including
overhead, profit, escalation and contingency). Divide the total
by the amount of your direct costs. This will give you a factor.
Multiply each bid item by that factor to find the bid price.

Here's an example. Figure 1-4 shows estimated direct costs,
overhead and profit for a sample job. Direct costs (materials,
labor and equipment) plus markup (overhead and profit) come

to $36,450.00. Divide this amount by the direct costs
($28,375.00) to get a factor of 1.2846. Now multiply each direct
cost item by this factor to find the bid price for each item.
Every cost line now includes your markup.

Item no.	Quantity	Price		Total
1	1000	@	$5.00	$ 5,000.00
2	2000	@	7.00	14,000.00
3	750	@	2.50	1,875.00
4	5000	@	1.50	7,500.00
		Total of direct costs		**$28,375.00**
Overhead and profit:	Office and shop			$ 2,837.00
	Supervision			1,600.00
	Mobilization			800.00
	Profit (10% of direct costs)			2,837.50
	Total of overhead and profit			**$ 8,075.00**
	Direct costs			$28,375.00
	Overhead and profit			8,075.00
	Total bid			**$36,450.00**

$$\frac{\$36,450.00}{\$28,375.00} = \text{Factor } \textbf{1.2846}$$

Item no. 1	1000	@	5.00 x 1.2846	=	$6.4223	=	$ 6,423.00
Item no. 2	2000	@	7.00 x 1.2846	=	$8.9922	=	$17,984.40
Item no. 3	750	@	2.50 x 1.2846	=	$3.2115	=	$ 2,408.63
Item no. 4	5000	@	1.50 x 1.2846	=	$1.9269	=	$ 9,634.50
							$36,450.53

Increasing bid items to cover overhead and profit
Figure 1-4

Here's a Simple Estimate
Every estimate begins with a labor, material and equipment quantity estimate. Cost items are listed one after another on a columnar pad like in Figure 1-5. Notice that these are direct job costs. No overhead or markup is included here.

Figure 1-5 shows the direct cost sheets for a slide correction project along a roadway. Note the 12 columns on each sheet:

- 1) Bid item number
- 2) Crew, equipment and materials
- 3) Quantity (hours or materials)
- 4, 5) Labor (rate and amount)
- 6, 7) Equipment operation (rate and amount)
- 8, 9) Materials (rate and amount)
- 10, 11) Subcontractors (rate and amount)
- 12) Totals

We enter the bid item number in the first column, followed by a brief description of the work in the second. The third column, quantity, is vital to making an accurate estimate. Enter here the estimated number of days, or feet of pipe, or cubic yards of rock, whatever unit of measure you're using for the particular item. Multiply this quantity by the *rate* for that bid component. In item 1, for example, we multiply 4 days labor by the labor rate of $160. Enter the total, $640, in the labor amount column and carry it over to the total column on the far right.

After calculating each item on the page, figure the total for the page at the bottom. To check your total, add up each "amount" column (labor, equipment, materials and subcontracts) at the bottom of the page. Total these amounts. You should arrive at the same answer by adding all of the figures in the total column and all of the figures at the bottom of the labor, equipment, materials and subcontract columns. If not, you've made a mistake. Go back and recheck all of your figures.

On the fourth page of Figure 1-5, we've totaled the direct costs from each of the first three pages and added supervision and bonding costs. The total estimated direct field costs for this job will be $34,203.60

If you have a computer, you can do this worksheet with a simple spreadsheet program. The program does the math for you, eliminating errors that creep in when you do it manually.

Direct Field Costs

Schedule No. _____   **Project** Road Slide   **Page** 1

Bid Item No(s). 1 thru 6   **Description** Remove slide, Install 12" drain and 6" filter drain   **Quantity** 4,500 yds. ex. 500' of pipe   **Est. By** D. Roberts   **Date** Sept. 7, 1984   B.O. Sept. 12, 1984

Bid Item #	Crew Equipment And Materials For Each Work Item	Quantity Also Hours Or Material	Labor Rate	Labor Amount	Equip. Operation Rate	Equip. Operation Amount	Construction Matls. & Supplies Rate	Construction Matls. & Supplies Amount	Subcontractors Rate	Subcontractors Amount	Total Rate	Total Amount
1	Clear & Grub:											
	2 days for 2 men w/saws	4 man days	160	640								640
	2 days for chainsaws	2 days			50	100						100
												740
2	Dozer:											
	Rental Rate	60 hrs.			80	4800						4800
	Dozer Operator	7 days	180	1260								1260
												6060
3	Mobilization:											
	Max. Allowable 6% of bid	100 miles			3	300						300
4	Compacted Road Base:	800 sq. yds.							3.50	2800		2800
5	Asphalt Surface:	800 sq. yds.							5.00	4000		4000
												6800
6	Concrete Catch Basin:											
	Subcontractor quote	$800 ea.								800		800
	Total This Page			1900		5200				7600		14700

Direct cost sheet

Schedule No. _____ Project Road Slide _____ Est. By _____

Bid Item No(s). 7 _____ Description _____ Quantity _____ Date _____

Bid Item #	Crew Equipment And Materials For Each Work Item	Quantity Also Hours Or Material	Labor Rate	Labor Amount	Equip. Operation Rate	Equip. Operation Amount	Construction Matls. & Supplies Rate	Construction Matls. & Supplies Amount	Subcontractors Rate	Subcontractors Amount	Total Rate	Total Amount
7	12" CSP w/Pipe Bedding:											
	Pipe	250 ft.					14.60	3650				3650
	Connecting band	12 ea.					14.00	168				168
	Pipe outlet	1 ea.					35.00	35				35
	Pipe bedding											
	250' x 2 x 2											
	———————— = 37 cu. yds.											
	27											
	wastage = 8 cu. yds.	45 cu. yds.					5.00	225				225
	Equipment											
	Backhoe	3 days			150	450						450
	Backhoe operator	3 days	180	540								540
	Backhoe laborer	3 days	160	480								480
	5548											5548
	250 ft. = 22.19 per ft. direct cost											
	Total This Page			1020		450		4078				5548

Direct cost sheet (continued)
Figure 1-5

Direct Field Costs

Schedule No. _____ Project __Road Slide__   Page __3__

Bid Item No(s). __8 & 9__   Description _____   Est. By _____

Quantity _____   Date _____

Bid Item #	Crew Equipment And Materials For Each Work Item	Quantity Also Hours Or Material	Labor Rate	Labor Amount	Equip. Operation Rate	Equip. Operation Amount	Construction Matls. & Supplies Rate	Construction Matls. & Supplies Amount	Subcontractors Rate	Subcontractors Amount	Total Rate	Total Amount
8	Filter Drain:											
	6# ADS perforated pipe	250 ft.					1	250				250
	Filter fabric											
	$\frac{250 \text{ ft} \times 15}{9} = 416$	416 sq. yds.					1	416				416
	1/4" drain chips											
	$\frac{250 \times 3 \times 4}{27} = 111$	111 cu. yds.					9.18	1019				1019
	Equipment											
	Backhoe	3 days			150	450						450
	Backhoe operator	3 days	180	540								540
	Backhoe laborer	3 days	160	480								480
	3155											3155
	250 ft = 12.62 per ft. direct cost											
9	Hydro Seeding:											
	Subcontractor's quote	1.5 acres								3000		3000
	Total This Page			1020		450		1685		3000		6155

Direct cost sheet (continued)
Figure 1-5

Schedule No. _____ Project Road Slide _____ Est. By _____

Bid Item No(s). _____ Description _____ Quantity _____ Date _____

Bid Item #	Crew Equipment And Materials For Each Work Item	Quantity Also Hours Or Material	Labor Rate	Labor Amount	Equip. Operation Rate	Equip. Operation Amount	Construction Matls. & Supplies Rate	Construction Matls. & Supplies Amount	Subcontractors Rate	Subcontractors Amount	Total Rate	Total Amount
	Total Direct Costs:											
	Total Page 1											1470 0
	Total Page 2											5548
	Total Page 3											6155
	Total Direct											26403.
	Other Costs:											
	Supervision	2 weeks	750	1500								1500 0
	Bond 2% of bid											600 0
	30,000 x .02 =											2850 3
											x	1.2
												3420 3.60

Direct cost sheet (continued)
Figure 1-5

But whether you use a computer, a ten-key or a calculator to do the computations, cross-check your answers as I described above. You can't afford math errors.

The estimate in Figure 1-5 is about as simple as they come. There's another more extensive estimate explained later in this chapter. But no matter how big or small the job may be, there are five important steps you'll take when making the estimate:

Step 1:   Examine the plans and specifications.

Step 2:   Visit the job site.

Step 3:   Prepare a tentative work schedule.

Step 4:   Secure bids from subcontractors.

Step 5:   Prepare direct cost sheets.

These direct cost sheets are the foundation for your bid. The quantities and prices listed must be accurate. If you don't have accurate numbers, don't guess. Do the work. Most bad estimators are lazy estimators. They cut corners to save time or effort. Don't be guilty of that crime. Call the suppliers. Talk to the subs. Get the opinion of your supervisors or equipment operators. Identify each work item in the job and list your expected cost. Check and recheck the plans for anything you might have omitted. Nail down each cost as accurately as humanly possible. The estimate isn't finished until there's nothing more you can do to make it a more accurate forecast of actual costs.

Avoiding and Finding Errors

The easiest way to prevent errors is to avoid them in the first place. The best insurance against errors is a consistent system, a procedure that starts with a careful review of the plans and specs and proceeds in some logical order from the beginning of the project to the end.

Here's the system I recommend: Visualize in your mind the first step in the project. Write down the name of that step on your direct cost estimate sheet. List all the direct cost components for that step: labor, material, equipment, or subcontract. Then go on to the next step in the project, identifying all direct costs for that step. Continue with all the following steps, writing down every cost item in the order the work will be done.

Leave plenty of space on your estimate sheet to do calculations and make changes. Show all your work. Don't write on the back of the sheet. You may forget to turn the sheet over when bringing page totals forward to an estimate summary.

When all work in the project is listed, begin extending quantities, estimating labor hours and pricing each work item.

If you've followed a logical order, used a checklist to catch errors and worked carefully, there won't be many errors. But even one error is too many. What do you do to get every last error out?

Recognize first that everyone makes mistakes. Estimators are no exception. Plan on it. You're going to forget key items, misplace decimal points, transpose numbers and worse. It's going to happen. But the mistakes you make won't cost you a dime if they're caught before the bid is submitted. That's why checking estimates is so important.

Check every estimate first by reviewing every line item and calculation. Never trust your results until you've checked all measurements and calculations. Addition *should* be done twice, but multiplications can often be checked by inspection.

Your own checking should identify 90% of the errors. What about the last 10%? Sometimes you can go over calculations again and again and still not find the error. You've become "blind" to the defect. That's why a second estimator is needed to verify your accuracy. Ideally the second estimator would make a second estimate without referring to your estimate, making completely new measurements and calculations. But there usually isn't time for that. Checking your work will be almost as good and lots faster.

Here are some of the most common errors to watch for when checking any estimate:

1) Errors in arithmetic such as in addition, subtraction, multiplication, division, and decimal points.

2) Errors in copying items from one sheet to another.

3) Omissions of items of materials, labor, subcontract or direct overhead. Use a checklist to be sure you've remembered everything.

4) Errors in estimating the manhours needed to do certain tasks.

Remember that anything that seems wrong probably is. Step back and ask yourself, "Is that figure about right?" If it seems too high or too low, keep checking. This is where bid tabs are useful. If a certain bid item usually bids for about $10 a linear foot, your estimated cost should be in that range. Your financial future and reputation in the construction community are on the line with every estimate. Give your estimates the care and attention they deserve.

When your estimate has been checked and verified, it's time to prepare the bid and submit it for acceptance. Transfer the estimate totals to a bid sheet, check to be sure the numbers were transferred correctly, put the bid and your bid bond in an envelope, and take it to the bid opening. Many estimators leave their bid unsealed until the last minute to allow for a sudden change of heart or late subcontract prices.

You've prepared the most accurate bid possible. If your numbers are good, you may win the contract and make a profit on the job. If you don't get the contract, don't worry. There will be other projects. Get ready for the next job. You never beat the guy who makes a mistake — but you don't want to.

Submitting Your Bid

The Associated General Contractors of America has published a booklet called *Bidding Procedures and Contract Awards,* which lays out the ground rules for competitive bidding. Order a copy from AGCA, 1957 E Street, NW, Washington, DC 20006.

Public agencies are required to award the contract to the "lowest responsible" bidder. They can ignore bids from irresponsible bidders such as a company that doesn't have the financial resources to carry the project.

I wouldn't advise that you to try guessing what the competition is going to bid. That's a waste of time. But you can usually guess who's going to bid against you. If a small job will attract too many bidders, stay away from it. The job will go cheap. Bidding costs you time and money. There's no advantage to winning contracts that won't earn a reasonable profit.

Some contractors front load (or unbalance) their bids by charging a higher rate for the work they're doing early in the project. This fattens progress payments received early in the project. The engineer will usually tolerate some front loading. But if you overdo it, he'll exercise his right to reject the bid. Figure 1-6 shows an obviously front-loaded bid.

There's another danger to front loading or distributing markup unevenly among all the bid items. What happens if you pad or load one particular item, and the owner decides to cut back on that item? You lose money. It cuts into your profits. Here's an example.

Let's say you're bidding a small job and your bid has only two items. Suppose you put your full markup on one item and bid the other item at cost. You'll earn your full profit on the job, provided that the job actually requires the quantities shown on your bid sheet. If there's an overrun on the item you've marked up, your profits increase. If the owner cuts back on this item, he's taking a chunk out of your profit.

Item 1)　50 units.
　　　　　Direct cost $1.00, 40% markup
　　　　　　　　　50 x $1.40 = $70.00
Item 2)　50 units.
　　　　　Direct cost $1.00, 0% markup
　　　　　　　　　50 x $1.00 = $50.00
　　　　　　　　　Total Bid: $120.00

The average markup on the two items is a fair 20%. But if the owner decided to decrease item 1 by 50% and increase item 2 by 50%, your markup would decrease to only 10%.

Item 1)　25 units @ $1.40　=　$35.00
Item 2)　75 units @ $1.00　=　$75.00
　　　　　　　Total　　　　　　$110.00

But the opposite can also be true. If you feel one quantity will be increased, loading that item can greatly increase job profit. This is a simplified example, of course, but it illustrates the point: unbalancing bids should be done with care. It's a dangerous game that can backfire by getting the bid rejected. By the same token, beware of an engineer who's playing a game with the bidders by grossly misstating quantities.

Item 1) Supply pipe, excavation, backfill   2,000 LF @ $ 10.00 = $20,000
Item 2) Supply fittings & install water meters   20 Ea @ $250.00 = $ 5,000
Item 3) Supply & install concrete curb & gutter   2,000 LF @ $ 3.00 = $ 6,000
Item 4) Asphalt restoration   2,000 LF @ $ 7.00 = $14,000

Total bid: $45,000

Loading item 1 by $5 per LF increases the total bid by $10,000. Subtract this $10,000 from items 2, 3 & 4 in proportion to their share of the rest of the bid. Simply taking the $10,000 from just one item would be very noticeable. Here's how to to it.

Total the unaltered items:

Item 2	$ 5,000
Item 3	$ 6,000
Item 4	$14,000
	$25,000

Divide each item by the total to find that item's percent of the total:

Item 2 $\dfrac{5,000}{25,000}$ = .2 x 100 = 20%

Item 3 $\dfrac{6,000}{25,000}$ = .24 x 100 = 24%

Item 4 $\dfrac{14,000}{25,000}$ = .56 x 100 = $\dfrac{56\%}{100\%}$

Multiply $10,000 by each of the percentages to determine the amount to reduce that item. Item 2's share is 20% of $10,000 ($2,000), which is subtracted from the $5,000 in the original bid. Find the revised unit prices by dividing the number of units into the adjusted totals:

Item 2) $5,000 - $2,000 = $3,000 ÷ 20 = $150.00 unit price

The final, front-loaded bid looks like this:
Item 1) Supply pipe, excavation, backfill   2,000 LF @ $ 15.00 = $30,000
Item 2) Supply fittings & install water meters   20 Ea @ $150.00 = $ 3,000
Item 3) Supply & install concrete curb & gutter   2,000 LF @ $ 1.80 = $ 3,600
Item 4) Asphalt restoration   2,000 LF @ $ 4.20 = $ 8,400

Total bid: $45,000

Front loading a bid
Figure 1-6

Sample Bid

Figure 1-7 shows a bid tabulation for a road slide correction project we bid on. Work through this estimate with me to be sure you understand the estimating process.

There were nine construction companies bidding on this job. Let's look at a description of the project. Then we'll prepare the direct cost sheets and the bid itself.

The work involves placing 250 feet of 12-inch pipe in a draw below the road. You'll install a catch basin at the upper end of the 12-inch pipeline and a 250-foot run of 6-inch perforated pipe, surrounded by 3/4-inch of gravel and wrapped in filter fabric. You'll remove the slide material by dozing it across the road and over the drain line. This work is intended to provide an escape for the water on the uphill side of the road.

The job is a little unusual because the slide removal has to be bid at an hourly unit price for a 120 to 140 horsepower dozer. The county will supervise the dozing work. Bidding the job at an hourly unit price allows the county to extend or reduce work without any complicated surveying.

Each bidder must bid on nine items:

1) Clear and grub 1.5 acres of light sagebrush.

2) Excavation and embankment. Rent a 120 to 140 hp dozer.

3) Mobilization. The specifications show that up to 6% of the total bid price can be spent on mobilization.

4) Resurface road with 800 square yards of compacted gravel base.

5) Finish road surface with 800 square yards of asphalt.

6) Install concrete catch basin at the intersection of the two pipelines.

7) Install 250 feet of 12-inch pipe with pipe bedding.

8) Install 250 feet of 6-inch perforated pipe with gravel filter and filter fabric.

9) Hydroseed 1.5 acres in the area of cut and fill.

Bid Abstract

Project Name __Slide correction__   Project No. _____   Sheet __1__ of __2__

Bid Date _____   Time _____   Place _____

Item No.	Item Description	Qty.	Unit	Smith Const. Unit Bid	Smith Const. Item Bid Totals	Jones Const. Unit Bid	Jones Const. Item Bid Totals	Brown Const. Unit Bid	Brown Const. Item Bid Totals
1	Clearing and grubbing	1	LS	2,000.00	2,000.00	2,300.00	2,300.00	1,920.00	1,920.00
2	Excavation and embankment	60	HR	80.00	4,800.00	90.00	5,400.00	80.00	4,800.00
3	Mobilization	1	LS	2,500.00	2,500.00	2,000.00	2,000.00	3,000.00	3,000.00
4	10" 3/4" - cr aggr roadway base matrl	800	SY	3.25	2,600.00	3.00	2,400.00	8.75	7,000.00
5	2-1/2" hot plnt mix asph conc surf cr	800	SY	5.60	4,480.00	6.00	4,800.00	6.48	5,184.00
6	Special drop inlet	1	EA	1,000.00	1,000.00	750.00	750.00	1,350.00	1,350.00
7	12" CSP 10 ga w/hugger bands	250	LF	25.00	6,250.00	20.25	5,062.50	20.00	5,000.00
8	6" corrugated/perf pipe underdrain	250	LF	13.00	3,250.00	10.25	2,562.50	12.00	3,000.00
9	Hydroseeding	1.5	AC	1,100.00	1,650.00	2,500.00	3,750.00	1,720.00	2,580.00
	Totals				28,530.00		29,025.00		33,834.00

Item No.	Item Description	Qty.	Unit	Green Const. Unit Bid	Green Const. Item Bid Totals	Reese Const. Unit Bid	Reese Const. Item Bid Totals	Nolan Const. Unit Bid	Nolan Const. Item Bid Totals
1	Clearing and grubbing	1	LS	2,500.00	2,500.00	4,000.00	4,000.00	4,785.00	4,785.00
2	Excavation and embankment	60	HR	90.00	5,400.00	120.00	7,200.00	70.00	4,200.00
3	Mobilization	1	LS	3,500.00	3,500.00	2,200.00	2,200.00	1,200.00	1,200.00
4	10" 3/4" - cr aggr roadway base matrl	800	SY	5.65	4,520.00	3.47	2,776.00	5.36	4,288.00
5	2-1/2" hot plnt mix asph conc surf cr	800	SY	6.50	5,200.00	4.81	3,848.00	6.42	5,136.00
6	Special drop inlet	1	EA	600.00	600.00	1,000.00	1,000.00	800.00	800.00
7	12" CSP 10 ga w/hugger bands	250	LF	20.36	5,090.00	29.09	7,272.50	23.84	5,960.00
8	6" corrugated/perf pipe underdrain	250	LF	15.00	3,750.00	17.58	4,395.00	19.80	4,950.00
9	Hydroseeding	1.5	AC	2,000.00	3,000.00	1,333.33	1,999.99	2,400.00	3,600.00
	Totals				33,560.00		34,691.49		34,919.00

Sample bid tab
Figure 1-7

Bid Abstract

Project Name Slide correction   **Project No.** _____   **Sheet** 2 **of** 2

Bid Date _____   **Time** _____   **Place** _____

Item No.	Item Description	Qty.	Unit	Bidder: Nelson Corp.		Bidder: White Const.		Bidder: Engineers Estimate	
				Unit Bid	Item Bid Totals	Unit Bid	Item Bid Totals	Unit Bid	Item Bid Totals
1	Clearing and grubbing	1	LS	6,000.00	6,000.00	2,000.00	2,000.00	1,500.00	1,500.00
2	Excavation and embankment	60	HR	200.00	12,000.00	100.00	6,000.00	75.00	4,500.00
3	Mobilization	1	LS	4,000.00	4,000.00	2,500.00	2,500.00	2,200.00	2,200.00
4	10" 3/4" cr aggr roadway base matrl	800	SY	4.16	3,328.00	10.00	8,000.00	4.00	3,200.00
5	2-1/2" hot plnt mix asph conc surf cr	800	SY	5.00	4,000.00	9.00	7,200.00	7.00	5,600.00
6	Special drop inlet	1	EA	1,200.00	1,200.00	4,000.00	4,000.00	1,000.00	1,000.00
7	12" CSP 10 ga w/hugger bands	250	LF	14.00	3,500.00	22.00	5,500.00	25.00	6,250.00
8	6" corrugated/perf pipe underdrain	250	LF	14.00	3,500.00	30.00	7,500.00	35.00	8,750.00
9	Hydroseeding	1.5	AC	666.67	1,000.00	4,000.00	6,000.00	200.00	300.00
	Totals				38,528.00		48,700.00		33,300.00

Sample bid tab (continued)
Figure 1-7

Preparing Your Direct Cost Sheets

Look again at Figure 1-5, the direct cost sheets for the road slide correction project. Earlier in this chapter I mentioned the five steps you'll take when preparing every bid. Let's review these steps for this bid.

Step 1:   Examine the plans and specifications.

Step 2:   Visit the job site.

Step 3:   Prepare a tentative work schedule. The work schedule must show both the equipment and the labor required to do each phase of the project. Your schedule for the road slide project will look something like this:

Days 1 & 2:   Clear and grub.   Two men with chainsaws.

Days 3 to 6:   Drive in wheeled backhoe. Excavate for catch basin. Install 12-inch pipe.   Backhoe, operator, laborer.

Days 7 to 12:   Move in dozer, doze slide material. Move out dozer.   Dozer and operator.

Days 13 to 15:   Install filter drain.   Backhoe, operator, laborer.

If you know your crew and are familiar with equipment production rates, you won't have any trouble coming up with a tentative work schedule. If you don't know your crew and you're not familiar with equipment production rates, get the best figures you can from other sources. Review your own cost records for the jobs you've just finished. Cross-check these figures with estimating tables and bid tabs.

Several major equipment manufacturers publish production rates for the equipment they sell. Review production rates for the equipment you plan to use. If you don't have any other source of information, talk to an experienced operator or superintendent. Form an opinion on how much work can be done in a day. The next chapter explains how to estimate labor and equipment productivity.

Step 4:   Get bids from subcontractors. The asphalt work, hydroseeding and concrete catch basin are all subcontract items on this job. Contact your subs and get their bids on these items.

Step 5: Prepare direct cost take-off sheets. Remember that your material calculations *must be accurate.* Chapter 2 explains how to calculate quantities.

Now let's take a detailed look at each of the nine bid items on the direct cost sheets shown in Figure 1-5.

1)   Clear and grub 1.5 acres of light sagebrush. For equipment, figure two days of chainsaw usage at $50.00 per day. For labor, figure two men operating the chainsaws for two days. This is the same as four days of labor at $160.00 per day.

2)   Excavation and embankment. For equipment, estimate 60 hours of dozer work at $80.00 per hour. For labor, estimate one operator for seven days at $180.00 per day.

3)   Mobilization. Estimate four trips of 25 miles each, for a total of 100 miles. The rate is $3.00 per loaded mile. On this job, the bid instructions allow only 6% of the total bid to be for mobilization.

4)   Resurface road. This is a subcontract item and is bid at cost. The sub quoted a rate of $3.50 per square yard for 800 square yards.

5)   Finish road. This is a subcontract item and is bid at cost. The sub quoted a rate of $5.00 per square yard for 800 square yards.

6)   Install catch basin. This is a subcontract item and is bid at cost. The sub quoted $800.00 for the catch basin.

7)   Install 12-inch pipe with pipe bedding. This item includes the pipe, pipe fittings, pipe bedding, and the labor and equipment required to install them.

Suppliers quoted on the pipe and pipe fittings. For pipe, estimate $14.60 per foot for 250 feet of pipe. For pipe fittings, estimate $14.00 each for 12 connecting bands and $35.00 for one pipe outlet.

Here's how to calculate the number of cubic yards of bedding required:

Multiply the trench length (in feet) by the trench width (in feet) by the bedding depth (in feet), and divide by 27.

For our road slide project, multiply the trench length (250 feet) by the trench width (2 feet) by the bedding depth (2 feet), and divide by 27. This gives us a total of 37 cubic yards of bedding required for this job. Notice that we've added 8 cubic yards for waste. If you're using small-diameter pipe, you can ignore the small amount of bedding displaced by the pipe. Just consider it waste.

For equipment, estimate three days of backhoe use at $150.00 per day. For labor, estimate one operator for three days at $180.00 per day and one laborer for three days at $160.00 per day.

8)   Install 6-inch pipe with filter and filter fabric. This item includes: the pipe, filter fabric, drain chips, and the labor and equipment to install them.

For pipe, estimate 250 feet at $1.00 per foot.

To figure the square yards of filter fabric required, multiply the length (250 feet) by the width (15 feet), and divide by 9 (square feet in a square yard). This gives you 416 square yards of filter fabric at $1.00 per square yard.

To figure the cubic yards of drain chips required, multiply the length (250 feet) by the width (3 feet) by the depth (3 feet), and divide by 27. The total is 111 cubic yards of drain chips at $9.18 per cubic yard.

For equipment, estimate three days of backhoe use at $150.00 per day. For labor, estimate one operator for three days at $180.00 per day and one laborer for three days at $160.00 per day.

9)   Hydroseeding. This is a subcontract item and is bid at cost. The sub quoted $3,000.00 for 1.5 acres of hydroseeding.

10)   Other costs, profit and overhead. Use the final page of your direct cost sheets to total up all your direct costs, add on other miscellaneous costs, and your percentage for profit and overhead.

Before totaling the direct costs, it's a good idea to round off all your unit prices to whole dollar numbers. This makes the final calculations easier and cuts down on errors.

Direct costs on the road slide project come to $26,403. Add on two weeks of supervision at $750 per week and a bid bond of 2% of the total bid. (Round the total bid up to $30,000 before you calculate the amount of the bid bond.) This brings our total costs to $28,503.

Now add on 20% for profit and overhead. You can spread markup evenly over all of the bid items or load all markup on a couple of items. If you're going to pad a couple of bid items, the mobilization and dozer rental rates are probably your best bets. But remember that you're taking a chance if you do this. The owner has control over the actual number of hours the dozer will be used on this job. If he wants to cut back on dozer work, it comes out of your pocket.

Bid Openings

Plan to be at the bid opening well before the deadline. At the appointed hour for opening you'll sit in suspense along with the other bidders. Each bid will be opened and read publicly. Every bidder wants the job, of course. There'll be a lot of tension in the air.

Bid openings can be very frustrating. If your bid is a great deal lower than the next lowest bid, the difference is money left "on the table." You'll be giving away some of your profit. If your bid is too high, a trimmed margin might have won the contract.

If your bid isn't lowest, you may still get the job if the lowest bidder isn't responsive to the invitation to bid. For example, the low bid may omit some items or be based on substitute materials. That gives you the right to raise objections. If there are valid objections to the lowest bid, the engineer may either award the contract to the next lowest bidder or throw out all bids and call for new bids. If the job is rebid, expect that all figures will come in at least a little lower than the original low bid. That's just simple human nature.

There's usually a delay between the bid opening and the awarding of the contract. If you're the low bidder on a large job, material suppliers will contact you during this interim period. Don't make any commitments. Tell them you're reviewing quotations and will give the order to the lowest bidder. You may want to invite quotes from other subcontractors too. But remember that it's not ethical, or smart, to reveal one supplier's (or sub's) quotes to another.

You're the Winner!

Once you've got a contract, review your estimate and talk to the people who'll be in charge on site. Ask your superintendent,

"This is how we bid the job — what do you think? What's the most cost-effective way of doing the project?"

Be especially careful when you review your equipment and material selections. The materials and equipment you use make a big difference in the final cost — and your profit. If alternate materials are allowed, check out every possible cost saving.

It's a good feeling to review a well-thought-out estimate and find that all your costs are covered. And it's even better when you find some cost savings.

After you've reviewed the project with your superintendent, you'll select your suppliers, subs and crews.

The next step is a preconstruction meeting with the engineer and owner. The engineer usually opens the meeting by expressing his requirements and making some general comments. Then it's up to you to ask any questions you have, and to provide the engineer with a work schedule. Don't leave this meeting without answers to all your questions. Serious problems or potential problems should be covered in written memo or letter form.

When the meeting is over, the fun begins. You and your superintendent will try to put the plan into action. If your estimate was accurate, there's a good chance that you can do the work for less than the bid price. In the next chapter I'll explain how that's done. But before we leave estimating, I want to cover two subjects that are important to every utility line estimator: cost keeping and bonding.

Controlling Costs

You can make two bankable profits on every job. The first is your estimated profit. The second is the extra profit earned by cutting costs below the estimate. Cutting costs isn't easy. But there are wasted time, wasted material, and inefficient use of equipment on every job. Your task is to find these and reduce them to a minimum. And the best way to find waste is with a good cost keeping system — a practical, detailed system of cost recording.

I'm not going to describe how to set up and run a cost keeping system. That would take more space than I'm willing to devote to this topic. Anyhow, several good books have been written on the subject. But I want to emphasize how important cost keeping is to your utility line contracting business. Every large, successful utility line contractor I know has some type of cost keeping system. I'm not sure they all grew and prospered just because they watched costs so closely. But all take the time and trouble to monitor costs on every job. If you want to run a prosperous, growing utility line contracting company, I'd advise you to do the same.

A contractor who has little or no payroll and who personally watches every part of every job may feel that he has complete control of all his costs. In fact, he may be right. But as his business grows, he has to adopt some system for controlling costs. No one can watch every part of every project when several jobs are going on at once in several parts of the county. No one can remember every detail of jobs that were finished months or years ago. The important cost facts have to be collected, organized and compared. That's what a cost keeping system does for you.

If you have good cost records, you can compare one job to the next, one crew to the next, and the productivity of one piece of equipment to the next. You can tell when things are going right, and make corrections when they aren't. You can do more of what works best and stop doing what doesn't work — long before it puts you out of business. You don't have to rely on hunches or opinions about what works best. You know. You've got the cost facts.

Contractors who don't keep cost records know something is wrong, because they're not making any money. But they may not discover exactly what's wrong until it's too late to fix it. Contractors with good cost keeping systems know something is wrong soon after it happens.

A cost keeping system is like a sensitive barometer. It shows change, the trend from day to day. It lets you compare conditions this week with conditions last week, month and year. By watching the barometer needle, you can anticipate trouble, finding out about potential problems while there's still time to avoid a major loss.

A cost keeping system has two purposes. First, it helps you reduce costs to a minimum. Second, it provides information on which you can base future estimates and make cost comparisons between different methods. To do this, the system has to collect information on your costs. The basic document for every cost keeping system is the daily production report your supervisor prepares. It should show the quantity of materials installed or work done, the working conditions (including weather), the crews used and hours expended, overtime authorized, materials and equipment used.

From these daily reports, you can determine the actual cost for each unit of work done — labor, material, equipment, subcontract and overhead for installing each foot of pipe, for example. Compare current costs to previous costs for similar work to identify waste (of material or supplies), inefficient labor, poor supervision, time lost (due to delays, equipment breakdowns, poor planning, or padded payrolls) and any of a hundred factors that can inflate your costs unnecessarily.

Cost keeping is like bookkeeping in some ways. But the two aren't the same. Cost keeping deals with costs on a unit basis. Bookkeeping deals with accounts, directly with profits or losses. Cost keeping is an engineering function. Bookkeeping is a clerical function. The cost keeping viewpoint is that of the engineer, dealing with unit costs and quantities of materials. The bookkeeping viewpoint is that of finance, dealing with cash balances and net profits. Because the two are so different, you shouldn't rely on a bookkeeper to record costs. It should be done by someone who's familiar with both construction and estimating.

That's all I'm going to say about cost record keeping. But I hope you're curious about this powerful cost-cutting, profit-building tool. And I hope you're aware that every growing utility line contracting business needs a cost keeping system to stay profitable.

Getting Bonds

If the project requires a bond and you can't get one, you're closed out. Utility line contractors who can't get bonds are left to scramble for the crumbs, while bondable contractors dine at a table spread with the more profitable work. That makes getting bonded one of your highest priorities if you want to handle public works projects.

Fortunately, there's a lot you can do to get the bonds you need. This section explains what your bonding company is looking for and how to meet their requirements.

Establishing a line of bonding credit can be your most important single step along the road to success in the utility line contracting business. Practically all public work is now bonded and more and more private work requires performance, labor and material bonds.

It's in your interest to develop a good working relationship with your bonding agent, a relationship based on confidence and cooperation. It should be much like the relationship you have with your bank. Your bonding agent puts together a presentation to the bonding company. It introduces your company, tells them who you are, what your experience in construction has been, and such information as:

1) Your personal financial condition.

2) Your company financial condition.

3) The names of material suppliers who extend credit to you.

4) A description of projects completed.

5) The names of personal references who know your reputation.

You can make the job of your bonding agent much easier by supplying all the material he needs to present your case. Letters from satisfied customers, photos of completed projects, and detailed resumes of key personnel are all considered in establishing your bonding capacity.

A typical cost for performance, labor and material bonds is between 1% and 2.5%. It varies among bonding companies and depending on the contractor's financial condition.

Some bonding companies write bonds at higher rates for contractors they consider at higher risk, either because of their financial condition or because of the size and type of work to be done.

Your bonding company may suggest that you not bid any more jobs for a while. It's usually wise to consider this advice carefully. Bonding companies know that a prime cause of contractor failure is taking on more work and larger contracts than you can handle. Also, remember that bonds are not like insurance. If you can't complete the job, the bonding company pays the loss. But they have the right to come after you for reimbursement. Your interest and the bonding company's are the same. They want to see you succeed as much as you do.

At one time or another you'll probably need the following types of bonds: bid bonds, performance bonds, payment bonds (labor, material and subcontract), combination performance and payment bonds, maintenance bonds, and sales tax bonds. Let's look at each of these bonds in detail.

Bid Bonds

Almost all public work requires that the bidder be bonded. This means that, along with your bid, you must submit either a certified check or a bid bond for a fixed percentage of the total bid.

The bid bond is the bonding company's guarantee that your bid is genuine, that you will enter into a contract if your proposal is accepted, and that you will furnish performance and payment bonds if granted the job. Your penalty for failure to do these things is either the total amount of the bid bond or the difference in price between the lowest bidder and the second lowest bidder, whichever is smaller.

Performance bonds assure completion of the contract according to its plans and specifications and within the time allowed. Payment bonds guarantee payment of labor and material bills.

Obviously, bid bonds and performance bonds go together. Requirements for the bid bond are the same as requirements for the performance bond. No bonding company issues a bid bond to a contractor unless they're also willing to issue a performance bond on the same job.

There's a dangerous trap here for the unwary. Bid bonds are usually small, sometimes only a few thousand dollars. Suppose you're in a hurry and include a cashier's check with your bid instead of a bid bond. Assume further that you get the job. Now comes the problem. What if no bonding company will give you a performance bond? You're stuck, and will probably lose the cashier's check submitted with the bid. That's why it's always best to qualify for the performance bond before bidding a job that requires one.

Even when either a certified check or a bid bond is acceptable, it may be best to use the bid bond. Otherwise, your money is tied up in outstanding bids. Most bonding companies provide bid bonds free!

You'll have to meet certain standards set by the surety company before being considered for any bond. I'll explain prequalification later. But once you've qualified, requests for bonds are considered on the basis of the merits of each contract.

On public contracts, both the amount and the form of the bond are specified by law. On private work, they are specified by the owner.

Performance Bonds
Once a qualified contractor is selected, the project is awarded
and the contract documents are signed. At this point, the
contractor has to provide a bond to guarantee his performance.
The performance bond guarantees that if the contractor fails to
complete the project, the surety company will get the job done
at no additional cost to the owner. Sometimes that's not easy.
Before issuing the bond, the bonding company will want to be
very sure of your finances and professionalism.

If you don't do the work, the bonding company has to finish
the job. But what if you claim you've finished, and the owner
claims you haven't? Then the question becomes "What did the
contract require?" In some cases, it's just doing the work
according to the plans and specifications. Other contracts may
require that you "promptly and faithfully perform the
contract." Or that you "faithfully perform all of the
undertakings, terms, conditions and agreements of the
contract."Although each of these phrases implies the same thing, the
courts have distinguished between them. Simple performance of
the work under contract is one thing. The last phrase
(sometimes called the *broad form*) requires performance of
every condition of the contract, many of which are not even
remotely related to your work.

Under this broad form contract language, your obligations
include *all* contract commitments. For example, you could be
held liable for a manufacturer's failure to meet warranty
provisions in the contract, or for failure to provide the required
insurance coverage. That's why the broad form bond requires
more study by your surety company.

Payment Bonds
An owner needs assurance that the contractor will pay for his
labor, materials and subcontracts. Any bill you don't pay
becomes a lien against the owner's property. Contracts require
that you pay all bills for labor, equipment, and materials used.
And that you discharge any liens filed by workers or suppliers
against the project. Payment bonds guarantee that all your bills
will be paid.

Combination Performance & Payment Bonds
Under this type of bond, there's a potential conflict between
claims of the owner and claims of suppliers of labor and

materials. For example, suppose there's a large loss. The face value of the bond may not be enough to satisfy the claims of both the owner and the material suppliers. That's why separate performance and payment bonds are usually used. They cost the same, or only slightly more.

Maintenance Bonds

Some of your contracts will require a one year warranty against defective workmanship and materials. This maintenance period guarantee is usually included in the performance bond. No additional bond is needed. But some owners request a separate maintenance bond. If the only warranty is against defective materials and workmanship for one year, there's no additional premium charge even if a separate maintenance bond is required. But if the warranty is broader, or for a term longer than one year, an additional premium is required.

Check the specifications for warranty provisions before you bid. And advise your agent if the coverage is beyond the minimum. That covers your extra risk and lets you include a maintenance premium cost in your bid.

Contracts to Avoid

Most of the contracts you sign are carefully drafted by government, municipal or private attorneys to guard the owner against every possible loss. Most of these protective clauses are in the general conditions. Many utility line contractors have been tripped up by what you'll find there.

Don't sign contracts that are so one-sided, no one could make a reasonable profit on the job. Read the general conditions carefully, and question your bonding agent about language you don't understand. There's no standard agreement for all construction work, though the American Institute of Architects forms are used by many architects and are generally considered fair.

If the contract is unfair, but can't be renegotiated, your bonding company may be able to write a tailor-made bond that provides more limited coverage. Since this doesn't help you solve your problem of an unfair contract, surety underwriters are reluctant to amend bond forms under these circumstances.

Your bonding agent won't bond on a contract that's clearly unfair or is poorly drafted. No responsible owner will demand that you sign a contract that is clearly unfair — especially if you point out the offending clauses. It's perfectly acceptable, and legally sufficient, to line through unfair contract clauses. Make sure you line out the same language on all copies, and have all parties initial the change.

What to Look For
Your bonding company will usually spot an unfair contract. They won't let you get in over your head. They protect themselves as much as they protect you. If the contract is unfair, maybe they can help you get the provisions changed so it's less one-sided. After all, if the owner requires a bond, but no bonding company will write one for the job, the work's never going to get done.

Even if your jobs don't require a bond, read and understand the paragraphs that follow. Unfair contracts have ruined many utility line contractors. Know what you're signing. Don't let some obscure sentence in a routine little contract wipe you out financially.

First, be sure the work to be done is described accurately. Normally, the contract will say that the work to be done is identified in the plans and specs. But the contract itself may require other work that isn't in the plans and specs. If you're bidding the plans and specs, it's easy to omit what's described only in the contract. Don't make that mistake.

Getting Paid
A fair contract provides progress payments for work completed. These payments should be based on a percentage of the work done to date. Generally, payments are made monthly. Payments should be based on a schedule of values for various portions of the work. If it's a unit price contract, payment will be based on quantities of work completed to date, as computed by the engineer.

An agreed-on percentage of each progress payment is withheld by the owner for his protection, until the job is completed and accepted. This retention is usually 10% of each progress payment. Better contracts provide that retainage is withheld from only the first half of the contract amount.

Contract provisions different from these may be unfair.

When you feel that the work is "substantially completed," there should be a procedure in the contract for getting paid. Usually you'll send the owner a notice of completion. Final payment should be due 30 days after the owner gets this notice. This 30 day period is called the lien period. Material suppliers, tradesmen and subcontractors have 30 days to file a lien if they haven't been paid. If no liens are filed, the owner should make the final payment.

Be sure that your contracts with subcontractors provide that they'll be paid only if, and when, you receive payment. That passes most of your risk of nonpayment to the subs. If there's a dispute between you and the owner and payment is withheld, you shouldn't have to pay your subs before you get paid.

Warranties

Your contract probably includes the following language: "Contractor warrants to the owner and the engineer all materials and equipment furnished under this contract will be new unless otherwise specified and that the work will be of good quality, free from faults and defects, and in conformance with the contract documents." The contract probably also requires that you remedy defects found by the owner "within one year after the date of substantial completion or within such longer period of time as may be prescribed by law or by the terms of any applicable special guarantee required by the contract documents."

These two quotes together are the basic warranty under most construction contracts. Your responsibility is limited to defective workmanship and materials, and to a reasonable period of time.

Warranties become unfair when they extend for an unreasonable time. Courts have held contractors liable for defects discovered many years after construction, but only for defects that weren't obvious or if fraud was involved. You aren't running an insurance company or a maintenance outfit. Your job is to do the construction. If what you do meets code, complies with the plans and specs, and no defects are discovered for 12 months, your liability should end. If the owner doesn't see it that way, maybe you shouldn't bid the job.

The basic warranty described above has been interpreted by the courts to require only that the work you do is suitable for the purpose intended. You have to install materials according to plans and do the work in a workmanlike fashion. You're not liable for damage resulting from errors and omissions in the plans. The engineer's mistakes aren't your responsibility. Yet engineers are often reluctant to take this responsibility. Beware of disclaimers engineers may add to an approval.

Also watch out for other warranties. They may be buried in technical specifications. For example, the contract may require that you provide a five year warranty on some piece of equipment. That's absurd! You're not in the equipment repair business. A fair contract would require only that you provide the owner with a written guarantee from the manufacturer.

Contract Duration, Delays and Damages
A fair contract states how long the job should take, either in calendar days or working days. It should also give acceptable reasons for extending the completion date:

1) Acts or omissions of the owner or architect

2) Delays caused by separate contractors employed by the owner

3) Delays caused by changes ordered in the work

4) Delays caused by labor disputes, fire or weather

5) Any delay beyond the control of the contractor

But some contracts offered by agencies of city and county government go on for pages trying to make you responsible for all delays that aren't entirely the owner's fault. That's clearly unfair. You should be responsible for your own delays and none other.

Many contracts provide for liquidated damages. You have to forfeit a certain amount for each day of unauthorized delay in completing a project. If work runs past the time limit, that amount is deducted from the final payment. If this amount is reasonable, the courts will probably enforce the contract provisions against you. That isn't necessarily unfair. In fact, it's to your advantage if it limits your liability for delay. But be sure the time period allowed is reasonable, and legitimate reasons for delay extend the completion time.

Don't take these provisions lightly. They can turn an otherwise attractive job into a sure loser.

Changes and Changed Conditions

The owner is normally given the right to add or delete work from the project during the construction period. Changes should be made only with a written change order. Never agree to make a change without a written authorization. You won't collect for changes made without the owner's written okay, especially if the owner and engineer aren't in complete agreement. Again, beware of engineer's disclaimer clauses. Look for something like this:

> *We approve the use of this material providing the contractor or manufacturer fully guarantee it.*

The A.I.A. contract, and most federal government contracts, spell out what will happen if *changed conditions* are encountered at the job site. You're entitled to extra pay for changed conditions. Conditions are considered changed, for example, when the soil type isn't what was indicated in the contract documents, or is not normal for the type of work you're doing.

Owners and engineers have written volumes of contract language, excusing themselves from liability for test borings and the other information they provide to bidders. Some contracts even say that you are responsible for conditions at the site, if those conditions aren't as indicated in the bidding documents. That's ridiculous. Be sure there's a "changed conditions clause" in the contract so you get paid if conditions aren't what the test borings showed, or are very unusual for the type of work being done.

Indemnity Provisions

Most of your contracts will include an indemnity provision that requires you to reimburse the owner for some losses. Suppose, for example, that one of your employees is driving your truck away from the job site, and runs down a pedestrian. You're responsible for the negligent acts of your employees. You'd be liable for the accident. There's nothing wrong with that.

Now, suppose the pedestrian sues the owner of the site and wins. By law, you aren't responsible for the owner's negligence. The pedestrian collects in full from the owner and you're off the hook. That's unfair. It was your truck that caused the damage in the first place, even though the accident happened at the owner's site.

It's precisely to remedy this kind of problem that owners, many years ago, began to require that contractors agree to reimburse them for certain losses. That's known as indemnity. Indemnification clauses in construction contracts make you assume the liability of the owner.

The "limited form" indemnification provision holds the owner harmless for your negligence or the negligence of your subcontractors. Nothing wrong with that. Unfortunately, things have gone far beyond this "limited form" of indemnification.

Even the A.I.A. indemnity clause is questionable, because it provides the owner with indemnity for losses due to joint negligence of the contractor and owner. Most contractors have come to accept this provision. It's now called the "intermediate form" of indemnification.

But there's no excuse for the "broad form" indemnification clause sometimes found in construction contracts. It puts you in the position of holding the owner harmless against losses due to the owner's sole negligence. That makes you an insurance company, insuring the owner, the engineer, and all their employees and agents. That's not your business. Don't sign a contract that includes a broad form indemnification provision.

Prequalifying for Bonds
Before you need your first bond, go to the insurance agent or carrier that handles your liability and workers' compensation insurance. Ask about getting construction bonds. If he doesn't handle bonds, he'll refer you to someone who does. If all else fails, the yellow pages of your phone book list bonding companies under "Bonds, Surety and Fidelity."

If you've never had a construction bond, the first step is to prequalify. Once you've prequalified, your requests for bid and performance bonds can be processed routinely. Plan to prequalify before you request the first bond. It may take several weeks to get your account set up. Allow plenty of time.

Getting prequalified is like getting acquainted. If you haven't worked together before and haven't been bonded before, your agent will have some basic questions:

1)   About your business. How long in business? Principal owner's experience? Form of organization? Key staff members and their experience?

2)   About the work you've done. Type of work completed? Geographical area where you operate? Size of jobs undertaken? Reputation with owners, architects and engineers?

3)   About the work you want to do. Type and size of projects? Number of jobs in progress at one time?

4)   About your reputation. With subcontractors? With suppliers? With other contractors? With your banker?

5)   About your history of meeting financial obligations.

6)   Do you get good professional advice? From your banker, accountant, lawyer, and insurance agent?

7)   Have you been successful?

The Basic Financial Documents
Here's where you get into financial statements — balance sheets and income statements. A balance sheet shows that you're solvent, and likely to remain solvent for the duration of the contract. An income statement (profit and loss statement, or P & L) shows the profit for the period covered. Both should be prepared by a certified public accountant (CPA).
Bond underwriters aren't like accountants. They don't follow rigid guidelines when making decisions about bonding capacity. They're more like gamblers. They're betting on your ability and professionalism. They're always looking for growing, profitable construction companies that will become good, steady customers. If they can see a bright future for you, they're anxious to do business with you. That's the good news.
The bad news is that you'll be judged primarily by the quality of your paperwork. Bond underwriters tend to think that

contractors make good decisions when all the information is in front of them — usually in written form. If you don't have good accounting records, if you can't supply a current balance sheet and P & L, if you don't have cost records for previous jobs, if you can't show a good, professional-quality estimate for the job you're bidding, they're going to be reluctant to write the bond you need.

Your bonding company will want you to use the "percentage of completion method" for reporting income. True, this calls for some estimating. But accurate estimating is your business. If you can't supply financial reports based on percentage of completion accounting, you're admitting that you have poor cost records, or little estimating ability.

Many contractors and accountants prefer the "completed contract" method of reporting income because it avoids paying tax on profits that may never be earned. But the percentage of completion method gives a better picture of your current condition, especially if completion percentage estimates are conservative. It also reduces the chance that you'll pay excess income tax.

Your "Financials"
Here are the documents your bond underwriter will want to see when you prequalify for bid and performance bonds:

1) CPA-prepared financial reports for the three most recent years. These should include:

> Balance sheets
> Income statements (P & L)
> Capital reconciliation
> The source and application of funds

2) The financial report should be audited, and be supported by an unqualified opinion by the CPA. This report must disclose the scope of the CPA's involvement.

3) The basis of reporting income should be identified.

4) A list of projects completed during the most recent year, and a schedule of jobs in progress showing profit estimates.

5)   A schedule showing the cost of owned equipment, depreciation of that equipment, and all liens against that equipment.

6)   A list of overhead expenses for the current year.

7)   A supporting statement describing any joint ventures reflected in the financial report.

8)   The financial report should include the accountant's footnotes.

Your bond underwriter will probably suggest that you keep his file of financial documents current, even if you aren't bidding more work right now. That makes it easier to process your bond requests. Sometimes, bid bonds can't be written because there isn't enough time to collect these basic documents. Avoid missing bid dates. Keep current financial records on file with your bond underwriter.

Other Information Needed
Besides these basic financial documents, your bonding company will want to know your:

1)   Bank accounts by bank and account number.

2)   Accounts receivable by name, address and date due.

3)   Notes receivable by name, address and date due.

4)   Inventory value.

5)   Stocks and bonds if listed on a national exchange.

6)   Real estate owned and rental income produced.

7)   Accounts payable by name, address and date due.

Unless your statement is prepared by a CPA and is certified, it will have to be verified either by phone or mail. That's why your financial statements have to be accurate and detailed.

When the prequalification process is complete, your bonding company will tell you the bonding capacity they recommend. The figure they give you will be a total for unfinished work on hand and, possibly, a maximum limit for any single job. This is your guide to the size of jobs to bid. Once your bonding limit is established, bid, performance, labor and material bonds will usually be issued very promptly.

To maintain your bonding account, keep your bonding agent up to date on bid results and the amount of work on hand. File new financial statements with your bonding agent at the end of each accounting year.

Bonds for Specific Jobs
Once you have a line of bonding credit, you'll usually be able to get the bonds you need, within your dollar limit and for the type of work you've been doing. But sometimes a bond request will be turned down because of some special risks in the contract. A difficult site problem, short working season, material shortages, restricted working conditions or natural hazards might discourage your bonding company. Or perhaps an unfair contract or labor-management relations problems will make the job too risky. Decisions like this are based on two judgments: First, that there's a serious risk. Second, that you can't afford to assume it.

Usually, when your bond underwriter turns you down, his explanation will be that the contract would leave you overextended. That means he feels you're biting off more than you can chew. For example, you're overextended if you take on too many jobs, take on jobs over too wide an area, take on work that you aren't used to handling, or simply take on too much work for your financial condition.

Some other factors can make the risk unreasonable in the eyes of your bond underwriter:

Owner financing— Doing business with some developers and promoters can be risky. If your owner runs out of money, you're not going to get paid. That can put a big hole in your bank account. Your bond underwriter will usually want to see money set aside in a special construction account for the benefit of contractors. This money should be paid directly to you, without actually passing through the owner's hands. This protects you against an underfinanced owner who might divert contract funds for other purposes.

The engineer— There's a risk in faulty plans and specifications and poor project design. An underwriter's judgment here will be only as valid as his knowledge of the engineering firms in his area. It's generally a good sign if you've done previous work with the engineers involved.

Subcontractors— You reduce your risk by subcontracting part of the work to responsible subs. The underwriter will be interested in the amount of work subcontracted and in the quality of the subcontractors selected. That's why your bonding company will want a list of subcontractors showing the name and location of the sub, the nature and amount of the work and whether or not a subcontract bond will be required. Sometimes your bond underwriter will require that major subcontractors post a bond. That's why you should check on the bonding capacity of your sub *before* using his bid.

Is the price right?— Your bonding company wants to be sure there's enough profit built into your bid. If there are other bids at about the same price as yours, that's good confirmation that your price is right. A spread greater than 10% between your bid and the next lowest bid may worry your bond underwriter. By then it's too late to turn down the bonding request, of course. But it's not too late to turn down your *next* request for a bid bond.

Strengthening Your Bonding Capacity
Your bond underwriter may not agree that the jobs you want are the jobs you should be handling. But don't give up without considering some other alternatives. I'll list the most common ways to make your bond underwriter reconsider:

Additional capital can be brought into the business, either by selling shares or by taking on a partner. This provides more cushion between the bonding company and a loss. But it isn't a substitute for management skill, and it doesn't eliminate risks in the job.

Subordination of debt gives the bonding company a prior claim against company assets if there's a loss. Subordination works when a small corporation owes money to its stockholders. The stockholders give the bonding company's claim priority over their claim. It's also used when a big debt is owed to a supplier.

Additional indemnity uses the pledge of some financial backer. He agrees to reimburse the bonding company if there's a loss on the bond. Getting additional indemnity makes sense if the financial backer has an interest in your company or stands to profit from the contract.

The joint venture makes sense on many jobs. Joint venture partners reduce the risk of failure by combining resources on a given project. If you enter into a joint venture, be sure the agreement covers the obligations of all partners on all important topics:

1) Advancing working capital

2) The percentage interest of each participant

3) Division of profits

4) Sharing of losses

5) Specific responsibilities and contributions

6) Details of administration and project management

7) Procedure in case of default by a partner

8) Termination of the agreement

A joint venture makes sense if your joint venture partner can add whatever is needed to make you qualify for a bond.

Here's Your Bonding Checklist

Follow the points listed below, and you should have no trouble qualifying for the bonds you need.

1) Keep up-to-date financial statements on file with your bonding company.

2) Have a certified public accountant audit your books and prepare your financial statements at least annually. Both your banker and bonding company need these documents.

3) Advise your bonding company as soon as you begin to figure a job. It takes time to underwrite even a bid bond. Try to give them an accurate breakdown of projected job costs. You can usually estimate your labor, material and equipment, but subcontractors may not give you a bid in time for the bid bond. Take care not to underestimate an unfamiliar subcontract item. Tell the bond agent that you're guessing on some items, and make sure you're conservative. It's better to ask for a $100,000 bid bond for a $60,000 job than be stuck with a $60,000 bid bond for a $100,000 job.

4) Never take on work that you don't have the reserves to carry. Don't count on profits from the current job until they're in the bank. Have enough cash available even if the current job doesn't earn a cent.

5) Keep your bonding company informed of progress on key jobs. Don't surprise them with bad news after it's ancient history. They deserve to be the first to know, not the last.

6) Get good professional advice. Be sure the quality of your legal advice is at least as good as the quality of your construction.

7) Set up and use an adequate cost keeping system.

8) Subcontracts shouldn't always go to the lowest bidder, nor material purchases to the lowest-priced supplier. Pick the sub who will help you earn the best profit and complete the contract promptly.

9) Don't waste cash on nonconstruction purchases. Your business is utility line contracting. Stick to it.

10) Many utility line contractors load themselves down with too much equipment. That cuts your financial strength. Keep your investment in equipment as low as possible. Never take work at no profit just so your equipment stays busy and cash keeps coming in. That's a prescription for disaster.

11)   Finally, remember what I said at the beginning of this chapter. Rome wasn't built in a day. Crawl before you walk, walk before you run. Go from small jobs to somewhat larger projects. Get your experience on little jobs where making a mistake won't bankrupt you. When you've built your financial muscle, move on to larger jobs. Be patient. You'll never join the ranks of the biggest contractors if your company goes belly up because you tried to expand too fast.

2

CREW AND EQUIPMENT PRODUCTIVITY

Every utility line job is unique. You'll never have two that are exactly the same. The job you're bidding today is different in a dozen subtle ways from any other job you've ever had, even if you're using the same equipment and installing the same materials. That tends to make even experienced utility line estimators humble — and give more than a few grey hairs before their time.

Estimating utility line crew and equipment productivity is an inexact science. There are many variables. It's hard to sort out what's important from what may only seem important. Any of several dozen variables can have a dramatic impact on production rates and profit. Other variables may be irrelevant.

But don't give up hope. We'll explore most of the important variables in this chapter — and offer some advice that will help you avoid the worst estimating mistakes.

The most important cost variables on every job will be:

- Equipment cost and performance

- Labor performance

If you do your take-off correctly, you're not going to make a big mistake when estimating material quantities. The difference between your highest and lowest material bid will probably be less than 10%. But selecting the wrong equipment, not using equipment to good advantage, or tolerating poor performance from your crews can cost you plenty, far more than 10% of your total job cost. That's what this chapter is about, predicting how much work your crews and equipment can do, and then making sure actual and estimated productivity match up.

Your own cost records will always be the best guide when estimating equipment and crew performance. That's why collecting and saving crew and equipment productivity data is so important. But what if you don't have cost records on previous jobs? You'll have to use published data, the kind of information you'll find in this chapter.

Before I begin on crew and equipment productivity, I'm going to explain how to calculate earthwork quantities. You have to know the units of material being moved before applying any production rate. The cost of excavation is the cost per unit times the number of units plus the cost of mobilization and demobilization. Mobilization is sometimes itemized as a separate cost on your bid. If it's not, you'll need to include it in your equipment cost calculations.

After we review how to calculate earthwork quantities, we'll look at optimum hourly production rates and the factors that affect these rates. Then we'll walk through three sample estimates. Finally, we'll consider how to compare equipment and crew costs to select the appropriate equipment and crew for each job you're bidding. All the calculations in this chapter can be done on an inexpensive hand-held calculator. If you don't already own one of these modern marvels, buy one. A ten-key that prints is better; a computer even better than that.

Volume of Material

Soil (and all fill material) is commonly measured in cubic yards. When you're measuring soil, you'll want to identify the *state* of the soil, as well as the number of cubic yards. Here are the three common soil states:

1) Bank cubic yards (B. CY): This measurement refers to soil resting in its natural, undisturbed state.

2) Loose cubic yards (L. CY): When soil is excavated or in some way disturbed from its natural state, it swells, increasing in volume.

3) Compacted cubic yards (C. CY): When soil is compacted, it shrinks, decreasing in volume.

Here's an example that shows how important it is to identify the state of the soil:

- 1,000 *bank* cubic yards of dry loam become 1,350 *loose* cubic yards when the soil is excavated.
- When you compact this 1,350 loose cubic yards, it becomes 880 *compacted* cubic yards.

You can see how important it is to allow for swell and shrinkage when calculating the volume of material. If you don't consider the state of the soil, your estimates can't be accurate.

Here's the formula to use when calculating the volume of material to be moved:

Multiply the length (in feet) by the width (in feet) by the depth (in feet) to get the number of cubic feet of material. Then divide by 27 to get the number of cubic yards.

Let's say you're digging a trench 100 feet long by 2 feet wide and 5 feet deep. Multiply the length (100') by the width (2') by the depth (5') to get 1,000 bank cubic feet. Now divide by 27 to get a total volume of 37.04 bank cubic yards of material.

As a practical matter, the depth of a trench will vary when pipe is laid across uneven ground. Trench depth may vary from 5 feet to 7 feet or even 10 feet deep. When the depth varies, it's acceptable estimating practice to figure the average depth and then base your volume calculations on that average.

Here's how to figure the average depth. From the plans, measure the depth at regular intervals along the ground where the trench will be dug. Then add the measurements together and divide by the number of measurements taken.

Assume you're digging a 100-foot trench. Trench width will be 2 feet. Measuring each 20 feet along the length of the trench, you find the depths will be 5, 5, 6, 7 and 10 feet. Add the depths together to get a total of 33 feet. Now divide by the number of measurements taken (5) to get an average depth of 6.6 feet. The more depth measurements you take, the more accurate your average depth will be.

$$5' + 5' + 6' + 7' + 10' = 33'$$

$$\frac{33'}{5} = 6.6' \text{ average depth}$$

Now let's calculate the volume of material to be moved from this trench. Multiply the length (100') by the width (2') by the average depth (6.6') to get 1,320 bank cubic feet. Divide by 27 to get a total volume of 48.89 bank cubic yards.

$$100' \text{ x } 2' \text{ x } 6.6' = 1,320 \text{ bank cubic feet}$$

$$\frac{1,320}{27} = 48.89 \text{ bank cubic yards}$$

Equipment Performance

Start the equipment production estimate by determining the maximum production rate for each piece of equipment. What's the most each piece of equipment can do under perfect conditions? Of course, it's highly unlikely you'll be working in perfect conditions. But once you know the optimum production rate, you can allow for the actual conditions and the operator's ability.

Hourly Production Rate

The optimum hourly production rate for each piece of equipment is the number of cycles per hour times the volume of material moved in each cycle.

**Optimum hourly production rate =
cycles per hour x volume per cycle**

Most major equipment manufacturers have spec sheets that show cycle times for their equipment. These cycle times are based on tests done under ideal conditions: level land, easy soil, no crosslines and an experienced operator. Production rates on these spec sheets are often somewhat optimistic, to say the least.

Figures 2-1A, 2-1B and 2-1C show sample cycle times for the wheeled backhoe, tracked excavator and wheeled loader. Use these cycle times as a guide. But remember, *there's no substitute for your own data based on actual observation.* Collect cycle time data from the equipment on your jobs whenever possible.

When you have a reasonable estimate of your equipment's cycle time in seconds or minutes, convert it to cycles per hour.

Wheeled backhoe, 14 foot hoe, 4 foot deep trench		
Description	**Time (seconds)**	**Soil type**
Fast	10	Hard dry loam
Average	12	Hard dry loam
Average	16	Hard pan or cemented gravel
Slow (hard digging)	20-30	Caliche

Wheeled backhoe, 14 foot hoe, 8 foot deep trench		
Description	**Time (seconds)**	**Soil type**
Fast	12	Hard dry loam
Average	14	Hard dry loam
Average	18	Hard pan or cemented gravel
Slow (hard digging)	20-30	Caliche

Cycle times for wheeled backhoe
Figure 2-1A

Tracked excavator, 120-150 hp engine, 20 feet maximum digging depth		
Description	**Time (seconds)**	**Conditions**
Fast	18	Trench depth less than 10' deep, good work room with few obstructions.
Average	22	Trench depth less than 12' deep, underground utilities present, small dump target.
Hard digging	30	Trench depth in excess of 12' deep, overhead obstructions.

Tracked excavator, 250-300 hp engine, 25-30 ft maximum digging depth		
Description	**Time (seconds)**	**Conditions**
Fast	20	Trench depth less than 10' deep, good work room with few obstructions.
Average	27	Trench depth less than 12' deep, underground utilities present, small dump target.
Hard digging	33	Trench depth in excess of 12' deep, overhead obstructions.

Cycle times for tracked excavator
Figure 2-1B

Wheeled loader, 4 yard loader		
Description	Time (seconds)	Conditions
Fast	30	Loading trucks on firm surface 70' travel distance
Average	35	Same as above.
Average	40 - 60	Backfilling trench, restricted work room, bucket loads dumped in small piles for easy leveling by compaction equipment.

Note: Where manufacturers give heaped bucket capacity, reduce the heaped capacity figures by 10%.

Cycle times for wheeled loader
Figure 2-1C

To convert cycle time in seconds to cycles per hour, divide 3600 (the number of seconds in an hour) by the cycle time. For example, the average cycle time for a tracked excavator during hard digging is 30 seconds, according to Figure 2-1B. Divide 3600 by 30 to find the cycle time of 120 cycles per hour. If your cycle time is in minutes, just divide 60 (the number of minutes in an hour) by the number of minutes in the cycle. If the cycle time is two minutes, there will be 30 cycles per hour.

Now you've got the number of cycles per hour. To calculate the hourly capacity, you also have to know the volume of material moved in each cycle. Volume per cycle varies with the equipment capacity and soil swell.

Equipment capacity factor— Most earth-moving equipment comes with spec sheets showing the equipment capacity. Bucket capacities are described as "heaped" or "struck." *Heaped capacity* is the bucket's maximum capacity when the material is heaped above the edges of the bucket. *Struck capacity* shows the contents of the bucket when it's filled to the level of the rim, with no material heaped over the rim. The struck capacity is the same as the water-carrying capacity of the bucket.

On equipment spec sheets, the capacities shown are often heaped capacities. You won't have fully heaped buckets on most cycles on a job. It's impossible to get a heaped bucket load when working in saturated sand and gravel or poorly shot rock. If the manufacturer's specs show heaped capacities, reduce the numbers by 10% before you apply any adjustment factor.

Material	Factor
Moist clay or loam	1.35 of struck capacity
Dry clay or loam	1.30 of struck capacity
Moist sand and gravel	1.17 of struck capacity
Well blasted rock	1.10 of struck capacity
Saturated sand and gravel	.95 of struck capacity
Poorly blasted rock	.95 of struck capacity

Bucket fill factors for backhoe
Figure 2-2A

Loose uncompacted materials	Factor	Factor
Loose, moist loam	1.1 of heaped capacity	1.32 of struck
Sand and gravel	1.05 of heaped capacity	1.26 of struck
Well blasted rock	.8 of heaped capacity	.96 of struck
Poorly blasted rock and slab	.6 of heaped capacity	.72 of struck

Note: As loaders engaged in utility work are usually working with loose materials, a load factor is not needed. However, if the backfill volume is calculated from in-place trench volume, it will be necessary to apply the appropriate load factor to convert from B. CY to L. CY.

Bucket fill factors for wheeled loader
Figure 2-2B

	Grade	Factor
Uphill (- grade)	20%	.66
	15%	.875
	10%	.916
	5%	.95
Downhill (+ grade)	5%	1.08
	10%	1.16
	15%	1.25
	20%	1.39
	30%	1.50
Slot dozing	25%	1.41
	30%	1.58
	35%	1.75
	40%	2.00

Blade capacity factors for dozer
Figure 2-2C

Figures 2-2A, 2-2B and 2-2C show equipment capacity factors for the backhoe, wheeled loader, and dozer. Use these factors to adjust your equipment performance estimate. For example, dry clay or loam has an adjustment factor of 1.30 in the backhoe table (Figure 2-2A). Multiply the struck bucket capacity claimed by the backhoe manufacturer by 1.30 to find the amount of material moved per cycle. Then multiply that number by the cycles per hour to find the optimum hourly production rate.

With the optimum hourly production rate in hand, it's time to look at soil swell.

Soil swell (load) factor— Factors that compensate for the swell in soil volume are called load factors. To find the load factor, divide the bank volume by the bank volume plus the swell percentage. Here's an example.

$$\text{Load factor} = \frac{\text{bank volume}}{\text{bank volume} + \text{swell \%}}$$

To adjust your optimum production rate to allow for soil swell, just multiply the hourly production rate by the load factor. Figure 2-3 shows the load factors for common swell percentages. Figure 2-4 shows swell percentages for different soil types. Remember, these are averages. Soil condition and degree of compaction affect these percentages considerably.

Say your optimum hourly production rate is 250 L. CY when working in soil with a 0.76 load factor. Multiply 250 by 0.76 to find an adjusted production rate of 190 B. CY.

Notice that when you apply the load factor, you convert *loose* cubic yards into *bank* cubic yards.

Swell (%)	Load factor	Swell (%)	Load factor
5	.952	35	.741
10	.909	40	.714
15	.870	45	.690
20	.833	50	.667
25	.800	55	.645
30	.769	60	.625

Load factors for common swell percentages
Figure 2-3

Material	Cubic yards, in cut-- weight (lbs)	Cubic yards loose		Cubic yards in fill	
		Percent swell	Weight (lbs)	Swell or shrink,%	Weight (lbs)
Basalt	4950	64	3020	36	3640
Caliche	2430	16	2100	-25	3200
Chalk	4060	50	2710	33	3050
Clay					
Dry	3220	35	2380	-10	3570
Damp	3350	40	2400	-10	3720
Concrete:					
Stone	3960	72	2310	33	2910
Conglomerate	3720	33	2800	-8	4030
Decomposed rock					
75% R, 25% E	4120	25	3300	-12	3700
50% R, 50% E	3750	29	2900	-5	3940
25% R, 75% E	3380	26	2660	-8	3680
Dolomite	4870	67	2910	43	3400
Earth, loam:					
Dry	3030	35	2240	-12	3520
Damp	3370	40	2400	-4	3520
Wet, mud	2940	0	2940	-20	3520
Earth-rock mixtures:					
75% E, 25% R	3380	26	2660	-8	3680
50% E, 50% R	3750	29	2900	-5	3940
25% E, 75% R	4120	25	3300	12	3700
Gravel, ave. graduation:					
Dry	3020	15	2610	-7	3240
Wet	3530	5	3350	-3	3640
Granite	4540	72	2640	33	3410
Gumbo:					
Dry	3230	50	2150	-10	3570
Wet	3350	67	2020	-10	3720
Lime	--	--	2220	--	--
Limestone	4380	63	2690	36	3220
Loam, earth:					
Dry	3030	35	2240	-12	3520
Damp	3370	40	2400	-4	3520
Wet, mud	2940	0	2940	-20	3520
Masonry, rubble	3920	67	2350	33	2950
Mud	2940	0	2940	-20	3520
Pavement:					
Asphalt	3240	50	1940	0	3240
Brick	4050	67	2430	33	3050
Concrete	3960	67	2370	33	2980
Macadam	2840	67	1700	0	2840
Peat	1180	33	890	--	--
Plyolite	4050	67	2420	33	3040
Riprap rock, average	4500	72	2610	43	3150
Rock-earth mixtures:					
75% R, 25% E	4120	25	3300	12	3700
50% R, 50% E	3750	29	2900	-5	3940
25% R, 75% E	3380	26	2660	-8	3680
Sand, ave. graduation:					
Dry	2880	11	2590	-11	3240
Wet	3090	5	3230	-11	3460
Sandstone	4070	61	2520	34	3030
Shale	4450	50	2970	33	3350
Silt	3240	36	2380	-17	3890
Slate	4500	77	2600	33	3890
Topsoil	2430	56	1620	-26	3280

Reproduced from Excavation Handbook, *by Horace Church © 1981 with permission of McGraw-Hill Book Company*

Swell factors
Figure 2-4

Operator ability factor— A highly skilled operator can handle equipment on virtually any type of terrain. He's equally at home in any type of soil, on any safe slope, and in even the most congested site. But you won't find many operators like that.

Most equipment operators have more limited skills. Every time the soil or job conditions change from ideal, production rates change and your costs increase. Most operators have average ability. Don't expect better than average production from average operators.

The skill level of most operators will be about 75% of maximum on most jobs. Some will do better and some will do worse. Until you know the skill level of your particular operator, use a correction factor of 0.75. Take your optimum hourly production rate, adjust it to allow for equipment capacity and soil swell, and then multiply the result by 0.75.

In the previous example, we reduced the production rate of 250 L. CY to 190 B. CY. Now assume the job is being done by an operator with average ability. Multiply 190 B. CY by the correction factor of 0.75. The hourly production rate is now down to 142.5 B. CY.

A highly skilled operator can get as much as 30% more production out of a piece of equipment than an operator with average skills. But don't count on superior production rates until you've seen your operator do superior work under varying conditions.

Human efficiency factor— No one works at peak productivity for an eight-hour day. But this factor doesn't necessarily imply that your crew isn't working efficiently. You'll lose some production time in refueling, maintenance and the usual construction delays.

Every job has minor and major holdups, not all of which are the fault of the operator. Pipe layers can only lay pipe after the trench is excavated. Backfill can't be placed until the pipe is laid. None of this can be done until the job is properly laid out and organized.

When you consider all the variables, 50 minutes of productive work per hour is probably the best you can expect on most jobs.

To find the human efficiency factor, divide the number of productive minutes per hour by 60. The factor for a 50-minute workhour looks like this:

$$\text{Efficiency factor} = \frac{\text{minutes worked per hour}}{60 \text{ minutes}}$$

$$\frac{50 \text{ minutes worked}}{60 \text{ minutes}} = .83 \text{ efficiency factor}$$

To continue our example, we'll multiply 142.5 B. CY by the human efficiency factor of 0.83. The realistic hourly production rate is 118.3 B. CY.

When a machine has to spend a lot of time waiting on trucks or pipe laying, actual production time may be only 30 minutes per hour. When that's true, the efficiency factor is 0.50.

Crew motivation has a major impact on crew efficiency. In fact, motivation may be the most important variable of all. That's why your superintendent is the key man. He provides the motivation. He must know the business from top to bottom and have the maturity and personality to get top production from his crew. He needs to know how to put the right people in the right positions and provide each crew member with the incentives he needs. The best operators and tradesmen are motivated by pride of accomplishment and a spirit of cooperation and teamwork. If there are some of these people on your crew, a good superintendent will massage egos occasionally.

Most people on your payroll probably aren't natural team players. They're independent individuals with their own values and opinions. Yet it's vital that they be molded into a productive team. A good superintendent does exactly that.

We've learned how to calculate our optimum hourly production rate. And we've seen how to adjust it to allow for equipment capacity, soil swell, operator ability and human efficiency.

Let's go over the production calculation process again, because it's one of the most important elements of estimating. You may not be able to precisely predict production rates, but you should be able to tell what's impossible in a given set of circumstances. You'll be in trouble if you base a bid on an overly optimistic projected production rate. The projected production rate is the basis of your bid, the premise on which you base your other calculations.

With experience, you can begin to "ballpark" production rates. Experience may show a certain size loader and skilled operator can load out 15 truckloads an hour. Experience may tell you you can get a trenching rate of 100 feet an hour from a certain backhoe-operator combination, or a pipe installation rate of 300 feet a day. Such proven production data is very valuable.

The danger lies in oversimplifying it. If a job was profitable at a footage installation rate of $10 a linear foot last year, that doesn't mean you can base a bid on those figures today. Unless today's job is comparable, that information is irrelevant.

The basis of a bid is how long will it take and how much will it cost. So use production calculations to estimate how long it will take. Then compare your estimates against your previous production records. Finally, compare them against previous bids of your competitors. If there's a wide difference between your projection and the competition, there's either a faulty premise, a miscalculation, or your competitors are going broke.

Now let's apply what we've learned. We'll estimate the equipment performance rates for the following equipment: a wheeled backhoe, a tracked excavator, wheeled loader and compactor.

Wheeled Backhoe

Our first sample production estimate is for a wheeled backhoe project, excavating for sewer service. The average depth of the trench is 9 feet. The average width is 2 feet. The length of the trench is 30 feet. This project is in a developed area, and there will be several cross-lines. Water, phone, power and gas lines run parallel to the sewer main.

The backhoe has to excavate the trench and supply pipe bedding from a nearby stockpile. A pipe layer follows with the service pipe. A follow-up backhoe (fitted with a trench compactor) will handle the backfill. The soil type is dry loam.

Volume of Material
Use our formula to calculate the volume of material. Multiply the length of the trench (30') by the width (2') by the depth (9')

to get 540 bank cubic feet. Divide by 27 to get a total volume of 20 bank cubic yards.

If you want to know how many bank cubic yards of material there are in each linear foot of trench, just divide the total volume (20 B. CY) by the length of the trench (30 LF). This comes to 0.66 B. CY per linear foot of trench.

Hourly Production Rate

Using the formula for calculating optimum hourly production rates, we'll multiply the number of cycles per hour by the volume of material moved in each cycle.

Look again at Figure 2-1A. The cycle time for an operator of average ability digging in hard, dry loam is 14 seconds per cycle. (To convert this to minutes per cycle, divide 14 seconds by the number of seconds in a minute. This comes to 0.23 minutes per cycle.)

$$\text{Minutes per cycle} = \frac{\text{seconds per cycle}}{60 \text{ seconds}}$$

$$\frac{14 \text{ seconds per cycle}}{60 \text{ seconds}} = .23 \text{ minutes per cycle}$$

To calculate the number of cycles per hour, divide 60 minutes by the number of minutes per cycle (0.23). This gives you 261 cycles per hour.

$$\text{Cycles per hour} = \frac{60 \text{ minutes}}{\text{minutes per cycle}}$$

$$\frac{60 \text{ minutes}}{.23 \text{ minutes per cycle}} = 261 \text{ cycles per hour}$$

Now let's figure the volume of material moved in each cycle. Look again at the bucket fill factors shown in Figure 2-2A. For dry loam, the bucket fill factor is 1.3. To find the bucket volume, multiply the bucket fill factor by the struck capacity. Struck capacity for this bucket is 6 loose cubic feet. Multiply 6 by 1.3 to get 7.8 loose cubic feet. Then divide by 27 to get the number of cubic yards. The volume of material in each bucket load is 0.29 loose cubic yards.

Now let's apply the load factor. The swell percentage for this type of soil is 35%. (See Figure 2-4.) Figure 2-3 tells us that 0.741 is the load factor for a 35% swell rate. Multiply the bucket volume (0.29 L. CY) by the load factor to get a more accurate load. The volume of material in each bucket will be 0.21 B. CY.

We've figured out the number of cycles per hour and the volume per cycle. Now we can calculate the hourly production rate. Multiply the number of cycles per hour (261) by the volume per cycle (0.21 B. CY) to get a production rate of 54.81 B. CY per hour.

Let's convert this hourly production rate by volume to an hourly rate per *linear foot*. This will tell us how many linear feet of trench we can dig in an hour. Here's the formula:

$$\text{Linear feet per hour} = \frac{\text{hourly production rate}}{\text{volume of material in each LF of trench}}$$

$$\frac{54.81 \text{ CY}}{.66 \text{ B. CY}} = 83.05 \text{ LF per hour}$$

Correction Factors

But this isn't our final estimate. We still have to allow for operator ability and human efficiency.

We'll assume an operator with average ability. That tells us to use the 0.75 factor. Multiply the number of linear feet per hour (83.05) by the factor (0.75) to get a new linear foot production rate of 62.29 LF per hour.

Now let's allow for human efficiency. Start with the 50-minute work hour and the correction factor of 0.83. Multiply the linear foot production rate (62.29) by 0.83 to get 51.70 LF per hour.

If you can dig 51.70 LF per hour, how long will it take to do the entire 30 feet of trench? Just divide the length of the trench (30') by the number of linear feet per hour (51.70) to get a total production time of 0.58 hour. To convert this to minutes, multiply 60 minutes by 0.58. The answer is 34.8. Round that to 35 minutes for this trench.

Total hours required to dig total length of trench $= \dfrac{\text{length of trench}}{\text{LF per hour}}$

$$\dfrac{30'}{51.70 \text{ LF per hour}} = .58 \text{ hour for the entire } 30'$$

.58 x 60 minutes = 34.8 minutes per trench

On this job, other delays should be included in your calculation. This will be a stop-and-start operation. The area around the riser pipe (or tee) in the main line will have to be excavated by hand. If a grademan works along with the equipment operator, production will move faster and there'll be less chance of damage. But there will still be considerable delay.

Cross-lines will slow production. If the soil allows easy digging, the delay will be about two or three minutes per crossline. If the soil is hard and the crosslines are hard to find, the delay may be five minutes per crossline. Hauling and placing the pipe bedding will also delay production.

Suppose the total of these delays comes to about 10 minutes. That makes the total production time for this task about 45 minutes.

Look for other factors that may affect your production rate. Street traffic, excavating under sidewalks, curbs or gutters, asphalt removal and avoiding damage to private property will all slow production. And don't forget to add time for startup in the morning and cleanup at the end of the workday.

Most utility line contractors try to avoid leaving a trench open overnight in a high traffic area. You may not want to start any new excavation near the end of the workday. Keep this in mind. If your crew is working an 8-hour day, you'll probably get only 7 or 7½ hours of productive work, not 8.

Tracked Excavator

Calculate trenching production for a tracked excavator the same way as for a wheeled backhoe — with one major difference. The cycle time will be longer because it takes more time to maneuver tracked equipment around obstructions like trees and power lines.

Figure 2-1B shows average cycle times for a tracked excavator under three typical working conditions. Use the procedures and formulas described earlier to determine the volume of material and hourly production rate for your project. Evaluate the delays you'll face, then calculate your final estimated trenching production.

On some jobs you'll use the tracked excavator for laying pipe or loading the spoil into trucks. After you've calculated the excavation time, add for delay caused by the pipe laying or truck loading operation. A tracked excavator may be a poor choice for loading trucks. The production rate is much lower, especially when trucks can't get close to the work, or when maneuvering room is restricted. It may be better to use a loader for spoil removal, and leave the excavator to excavate.

Production will be delayed further when a tracked excavator is used to provide pipe bedding. On some jobs a tracked excavator has to be used because a wheeled loader won't fit in the work space.

Wheeled Loader

On a loading job you'll have four separate operations: excavation, loading, hauling and backfilling. Ideally, the production rate for the four operations match perfectly and all equipment works at top efficiency for the project duration. I've never seen a job like that. You won't either. Usually one operation sets the production pace. The others work at less than peak efficiency.

Loader capacity usually sets the work pace and dictates the number of trucks needed in a hauling job. In utility work, excavating equipment usually sets the work rate. Select backfilling equipment that can keep pace with other equipment.

Loader Bucket Capacity

If a tracked backhoe generates 100 loose cubic yards of spoil per hour, you'll need a loader that can handle 100 L. CY per hour. Here's the formula that tells you the loader capacity needed to keep up with the excavator:

$$\frac{\text{L. CY per hour}}{\text{Corrected cycles per hour}} = \text{bucket capacity required}$$

To find *corrected cycle times per hour,* first look up the optimum cycle time in Figure 2-1C. Convert this to optimum cycle times per hour: Divide the number of seconds in an hour by the number of seconds per cycle. If the average cycle time is 35 seconds, here's your calculation:

$$\frac{\text{Seconds per hour}}{\text{Seconds per cycle}}$$

$$\frac{3,600}{35} = 102 \text{ optimum cycles per hour}$$

Now we'll convert this from optimum to realistic cycles per hour. Apply the efficiency factor (0.83) and the operator ability factor (0.75).

$$102 \times .83 = 84.66 \times .75 = 63 \text{ corrected cycles per hour}$$

We're dealing with loose cubic yards, so we don't have to convert from bank to loose measure in this case. Now calculate the bucket capacity needed. Divide loose cubic yards per hour by cycles per hour. Our calculation looks like this:

$$\frac{100 \text{ L. CY per hour}}{63 \text{ cycles per hour}} = 1.58 \text{ L. CY bucket capacity}$$

To keep pace with the excavation rate, the loader bucket must carry 1.58 loose cubic yards per cycle. Of course, loader bucket capacity depends on the type of material being handled. Look back at Figure 2-2B. Use this formula to convert the specified bucket capacity to the appropriate type of material (in this case, moist loam):

$$\frac{\text{Specified bucket capacity}}{\text{Bucket fill factor}} = \text{required bucket capacity}$$

$$\frac{1.58 \text{ L. CY}}{1.1 \text{ (moist loam)}} = 1.43 \text{ L. CY heaped capacity}$$

If the material is well-blasted rock instead of loam, use the bucket fill factor of 0.8:

$$\frac{1.58 \text{ L. CY}}{.8} = 1.97 \text{ L. CY heaped capacity}$$

Leave a margin of safety. When you have to make a choice, select a bigger loader with extra bucket capacity rather than a smaller loader barely big enough for the task. That increases the loader cost but will reduce overall job cost.

Loader Production
Once you know the type of soil, the corrected bucket capacity and the cycle times, calculate loader production with this formula:

Loader production = cycles per hour x volume per cycle

Using the figures we've calculated above, here's the loader production for our project:

Loader production = 63 cycles per hour x 1.97 L. CY

Loader production = 124 L. CY per hour

Cycle Times
Figure 2-1C shows preliminary loader cycle times. Use this table only if you don't have better information from actual observations in the field.

The two main variables affecting loader production are travel distance and bucket dump time. If bucket dumping requires precise placement of spoil, or multiple dump targets, the cycle time will be longer.

When the loader has to carry loads over long distances, you may have to calculate cycle times from the four components of a loader's cycle: loading, traveling, dumping, and returning to load.

Loading the bucket will usually take about 6 to 10 seconds. Dumping the bucket can range from 3 to 5 seconds to as much as 20 seconds when the loader must dump in several piles.

Travel speed depends on the terrain and gears of the machine. Two miles per hour is an average loaded travel speed over rough or steep terrain. Four to eight miles per hour may be possible on smooth surfaces with an empty bucket. Use this formula to figure travel time:

$$\text{Travel time in minutes} = \frac{\text{distance traveled in feet}}{\text{speed per minute in feet/minute}}$$

To convert speed in miles per hour to feet per minute, multiply by 88. For example, 2 miles per hour times 88 equals 176 feet per minute.

Four steps are needed to calculate loader cycle times for longer carries. For example, here's the calculation for a 200-foot cycle distance:

1) Travel time   1.14 minutes

$$\frac{200 \text{ feet}}{176 \text{ feet per minute}} = 1.14 \text{ minutes}$$

2) Return to load at 4 mph   .56 minutes

$$\frac{200 \text{ feet}}{352 \text{ feet per minute}} = .56 \text{ minutes}$$

3) Bucket loading time (7 seconds)   .11 minutes

4) Bucket dumping time (3 seconds)   .05 minutes

                                    Cycle time   1.86 minutes

Bulldozer

Although few manufacturers list bulldozer blade capacity, several formulas can be used to get approximate figures. Here's the formula I prefer:

Height of blade in feet squared x width in feet
x .022 = loose cubic yards of blade capacity

Using this formula, we can figure the capacity of a dozer equipped with a blade 3 feet high and 10 feet wide:

$$3^2 \times 10' \times .022 = 1.98 \text{ L. CY blade capacity}$$

Grades affect both dozer pushing power and blade capacity. Figure 2-2C shows approximate blade capacity correction factors. Slot dozing, in which a dozer pushes material in a slot the width of its blade, is a very efficient method of moving material short distances downgrade. The slot increases blade capacity and the downgrade increases pushing power.

There are also several formulas for calculating dozer pushing power. Unfortunately, they're too complex for practical application. I prefer the following rule of thumb: *Assume that a dozer will push a blade load up a 20% grade at a speed slightly less than its maximum travel speed in first gear.*

Dozing speeds range from 1 mile per hour (88 feet per minute) up to 4 miles per hour (352 feet per minute). But it's hard for the operator to adjust blade depth and tilt fast enough to keep pace at high travel speed. Most dozers have a fast travel speed of 6 to 7 miles per hour. This is strictly a travel speed, not a working speed. But the fast travel speed can be used when reversing for the next dozing pass.

Here's another rule of thumb: *Average dozing speed is 2 miles per hour. Average travel speed when the machine is not under load will be 4 to 5 miles per hour.*

Bulldozer Cycle Time
A dozer's cycle time is made up of travel time forward under load plus the return time. The return time is usually about half of the travel time under load. In restricted work areas, however, the operator will spend more time maneuvering into position. Shorter dozing distances and confined work areas reduce travel speeds.

Dozing Distances
Obviously, a dozer moving earth an average of 100 feet will have a higher production rate than a similar machine moving earth 200 feet. Of course, more travel time is involved. Another reason is that soil tends to drift away from the blade as the

Dozer HP	Blade	Average distance (in feet)
50- 70	6'- 8' W, straight blade	75
70-150	8'-12' W, straight blade	100
150	12' W, semi-U blade	125
More than 150	U or semi-U blade	150

Note: Hourly L. CY production rates decrease by approximately 20 % per 100 LF beyond efficient average dozing distance.

Efficient dozing distances
Figure 2-5

dozer moves. This reduces blade load at the end of the pass.

On a long dozer push, use one of the special purpose U-shaped blades that are available for large machines. As a rule, the semi-U blade fitted to a midsized dozer increases capacity about 40% over the capacity of a straight blade. The full-U blade has a capacity about 1.2 times the capacity of the semi-U blade. Small utility dozers (50 to 150 horsepower with blade capacities from 1.5 to 4 loose cubic yards) can be fitted with straight multipurpose angling blades. These machines aren't a good choice for long distance dozing.

Figure 2-5 shows practical dozing distances for dozers. If material must be pushed beyond these average distances, use loaders and trucks or scrapers.

Estimating Dozer Production

To calculate optimum dozer production rates, use this formula:

Cycles per hour x L. CY per cycle = optimum dozer production

Here's an example. Suppose you need to backfill a 15-foot deep trench in an open area. The average trench width is 4 feet. The average dozing distance is about 50 feet. Assume a dozer speed of 1 mile per hour (88 feet per minute) under load, and 3 miles per hour (264 feet per minute) on the return. You'll use a 70 horsepower dozer equipped with a blade 3'6'' high and 8' wide.

    1) First, figure the amount of material in the trench:
        5 yards (15') x 1.33 yard (4') = 6.65 B. CY

2) Convert to loose cubic yards: (35% swell = load factor .74)

$$\frac{6.65 \text{ B. CY}}{.74} = 8.98 \text{ L. CY per linear foot}$$

3) Figure the cycle time and cycles per hour:

Travel time forward: $\dfrac{50'}{88 \text{ ft/min}}$ = .57 minutes

Return pass time: $\dfrac{50'}{264 \text{ ft/min}}$ = .19 minutes

Cycle time   .76 minutes

Cycles per hour: $\dfrac{60}{.76}$ = 79 cycles per hour

4) Calculate blade capacity:
$3.5^2 \times 8 \times .022 = 2.16$ L. CY per cycle

5) Figure optimum dozer production:
79 cycles per hour x 2.16 L. CY = 171 L. CY per hour

6) Correct for efficiency and operator ability factors:
171 L. CY x .83 (50 min. hr.) x .75 = 106 L. CY per hour

7) Convert hourly production rates in L. CY to linear feet per hour:

$$\frac{106 \text{ L. CY per hour}}{8.98 \text{ L. CY per linear foot}} = 11.8 \text{ LF per hour}$$

Figure 2-6 is a dozer production table. Use it as a guide unless you have more accurate information. The variables in your project can make a big difference. If you do use the table, be sure to read the notes and make adjustments for soil type and operator ability where appropriate.

	Net HP	Blade size in feet		Capacity in L. CY	Distance in feet	Cycles per hour	L. CY per hour (flat)	L.CY per hour (on grade)			
		Height	Width					-20%	-30%	-40%	+20%
Straight blade	80	3.0	8	1.6	50	85	136	166	217	272	108
	120	4.0	10	3.5	75	65	227	295	363	454	181
	160	4.5	12	5.4	100	50	270	351	432	540	216
	230	5.5	14	9.4	150	35	329	427	526	658	263
	350	6.5	15	14.0	250	21	294	382	470	588	235
Semi-U blade	120	4.0	10	4.9	75	60	294	382	470	588	235
	160	4.5	12	7.5	100	50	375	487	600	750	300
	230	5.5	14	13.0	150	35	455	591	728	910	364
	350	6.5	15	19.6	250	21	411	535	658	822	328
Full U blade	160	4.5	12	9.0	100	50	450	585	720	900	360
	230	5.5	14	15.8	150	35	553	719	885	1,106	442
	350	6.5	15	23.5	250	21	493	640	788	986	394

Notes: Figures are approximations based on dozing speed of 1.25 mph, reverse travel speed of 6 mph, easy loading soil and operator of above average ability.

On grades: – % = favorable downgrade dozing. + % = unfavorable upgrade dozing.

Optimum hourly production rates - dozer
Figure 2-6

Compaction Equipment

Chapter 6 covers compaction in detail. It's always hard to predict compaction equipment output accurately. Every soil type has unique compaction requirements and different shrinkage characteristics as it changes from loose to compacted condition. Refer back to Figure 2-4 for shrinkage factors for different soil types.

To determine the compacted volume of the material you're working with, use this formula:

Bank volume — shrinkage % = compacted CY volume

For example, caliche shrinks 25% after compaction.

1,000 B. CY — 25% (250 CY) = 750 compacted cubic yards

Two machines are commonly used for trench compaction: the sheepsfoot roller and the hoe-mounted vibratory compactor. We'll examine production rates for both.

Sheepsfoot Roller

The sheepsfoot roller is a good choice if you have a wide trench and no crosslines. The formula for calculating compaction output of a sheepsfoot roller is:

$$\frac{\text{Width of roller} \times \text{speed in mph} \times \text{thickness of compacted lift} \times 16.3}{\text{Number of passes necessary for compaction}} =$$

Maximum hourly production in compacted cubic yards

For the sheepsfoot roller, use an efficiency factor of 0.83 and an operator ability correction factor of 0.9.

For example, suppose a 5-foot wide sheepsfoot roller is working at 1 mile per hour on 12-inch layers (called *lifts*) of loose material. Each lift shrinks to 7 inches after compaction. Six passes are required.

$$\frac{5' \times 1 \text{ mph} \times 7'' \times 16.3}{6 \text{ passes}} = \frac{570}{6} =$$

95 compacted cubic yards per hour

95 C. CY x .83 x .9 = 71 C. CY corrected hourly production

This is a good first approximation. But only experience in the field with the soil involved can tell you the actual number of passes needed to compact soil to the required density. It's good practice to overestimate the compaction equipment and time needed. Don't get caught short with an overly optimistic estimate.

Hoe-mounted Vibratory Compactor

The hoe-mounted plate is ideal for narrow trenches and for compacting around crosslines. But estimating production of a hoe-mounted compaction plate isn't any easier than estimating sheepsfoot roller production.

In easily compacted soils, three to five seconds of pressure on a one-foot lift will give adequate compaction. This time can double in difficult soils. If you can't get adequate compaction in 10 seconds, then something's wrong. Either the soil is above or below optimum moisture, or the machine is too light for the soil being compacted.

For hoe-mounted plates, assume an efficiency factor of 0.83 and an operator ability factor of 0.8.

Here's the formula for calculating optimum production of a hoe-mounted plate.

$$\frac{\textbf{Area of plate in SF} \times \textbf{thickness of lift in F} \div 27 \times 3{,}600}{\textbf{seconds per position (including repositioning time)}} = \textbf{optimum production}$$

In this example, we'll assume a 2' by 2'6'' plate, 7'' compacted lift (0.58 feet), and 7 seconds per position, including repositioning time.

2' x 2.5' x .58' ÷ 27 x 514 (3,600 ÷ 7) = 55.2 C. CY per hour

55.2 x .83 x .8 = 36.6 C. CY corrected hourly production

Comparing Crew and Equipment Costs

Spend some time selecting the right equipment. It's an important decision. The best way to be sure you've got the right equipment is make up an estimate for each type that could be used. That's what we're going to do in the remainder of this chapter.

Selecting the right combination of crew and equipment is the key to profits for every utility line contractor.

Before making your final estimate, compare direct costs for the most likely equipment combinations. If using bigger equipment and a larger crew will double the production rate while increasing the daily cost only 50%, do it. But bigger isn't always better. Larger crews and more or bigger equipment won't always boost the work rate enough to make up for the increased cost. There's no advantage to using a $200-an-hour machine if the work can be done as easily with a $40-an-hour machine.

Think in terms of balances and trade-offs. A higher daily cost will often pay off in lower unit costs. But not always.

The daily operating cost of a piece of equipment is only part of the picture. Don't fall into the trap of assuming that the equipment you own is right for every project. That's ignoring reality. You're most competitive when you're using the right equipment. When appropriate, rent your equipment out to someone who can use it productively while you use a rented machine to do what you need to get done.

Most utility line contractors are fairly conservative and very cost-conscious. But you can increase efficiency and profits by finding new and better ways to get the job done. Be open-minded. Look at all the options and alternatives before you decide.

Think of earth moving as volumes of material to be moved in certain time spans and at a certain dollar cost. When you change the volume of material or the time allowed to move it, you change the cost. While it's almost impossible to estimate precise production rates, it's fairly easy to estimate *maximum* rates. Compare the calculated maximum with your actual production in the field. Chances are you'll be shocked at how much actual production is below what you believe is possible.

For example, watch truck loading with a tracked excavator. If your stopwatch shows 15-second cycles and it takes 12 cycles to load a truck, the trucks could in theory be loaded in three minutes. But actual loading will take longer. Figure out why.

There will always be delays as each truck moves into place. Is the truck driver positioning the truck poorly or too slowly? Is the operator slowing production with poor bucket loads or slow cycles? Is the material hard to load? Identify the problems that are lowering production rates on your jobs. Then either allow for them in production estimates or take steps to correct them.

When you have all the data to make your production estimates as accurate as possible, you're ready to compare production for various crew and equipment combinations. If your figures are accurate, it will be easy to pick the best combination.

Volume of Material
Let's look at an example. We'll compare crew and equipment options for an urban storm pipeline 1235 feet long and 2 feet wide. There are quite a few crosslines, but soil conditions are relatively good.

The first step is to find the cubic yards to be excavated. Calculate the volume of material from the plans. Average the trench depth by taking measurements on the plans each 200 feet along the line. Scale off the distance from the surface to the bottom of the trench. Remember, the formula is length times width times depth divided by 27. The answer is in bank cubic yards. Here are your calculations:

Depth measurements:

1	3.5'
2	3.5'
3	5.0'
4	7.0'
5	6.5'
6	4.5'
7	4.5'

34.5' ÷ 7 = 4.92' average depth

4.92' (depth) x 2' (width) x 1,235 (length) = 12,152.40 B. CF

$$\begin{array}{ll} 12,152.40 & \text{B. CF} \\ +3,038,10 & \text{(25\% swell)} \\ \hline 15,190.50 & \text{L. CF} \div 27 = 562.61 \text{ L. CY in the trench} \end{array}$$

$$\frac{563 \text{ L. CY}}{1,235} = .46 \text{ L. CY per linear foot}$$

Notice that this estimate ignores the extra excavation that's needed around the manholes. You could calculate manhole excavation volume from the dimensions on the plans. In our sample estimate, however, the volume of the manhole excavation is insignificant. We can safely ignore it. But if the work was for a deep sewer in poor soil, manhole excavation and shoring could be a major cost — too much to be ignored. It would be worth your time to make these calculations.

Once the volume is figured, think through the job step by step. First you'll have to remove the asphalt. It's usually best to do this before trenching begins. Removing it during trenching slows the work. If you excavate both at once, asphalt and spoil have to be piled separately. That takes up more space than may be available on a busy street.

On this project, you'll cut approximately 2,500 linear feet. That's enough to make a mechanical cutter economical. A good choice in this case would be a cutting wheel mounted on a grader, loader or backhoe.

You decide to price the asphalt removal as a separate item. Add this cost as a unit price after you've selected the best equipment and crew combination.

Now you can calculate production rates for crew and equipment combinations that could be used on this job. Follow the steps described earlier in this chapter. Find the optimum production rate for each piece of equipment. Reduce optimum rates to realistic rates by applying correction factors. Use 0.75 for operator ability, 0.83 for efficiency, and 0.6 for site conditions, including an allowance for crosslines.

Estimating Production for Wheeled and Tracked Backhoes
Begin with the wheeled backhoe. Suppose your observations indicate that the following rates should be used in these calculations: Average bucket load, 7 cubic feet. Average digging cycle time, 12 seconds (5 cycles per minute).

7 CF x 5 cycles = 35 L. CF per minute
35 L. CF x 60 minutes = 2,100 L. CF per hour

$$\frac{2,100 \text{ L. CF}}{27} = 77 \text{ L. CY per hour optimum production}$$

Reduce this to a realistic production rate by applying the correction factors:

$$77 \text{ L. CY} \times .75 \times .83 \times .6 = 28 \text{ L. CY per hour}$$

$$\frac{28 \text{ L. CY}}{.45 \text{ L. CY per LF}} = 62 \text{ LF per hour}$$

Now you have estimated linear feet per hour production for a wheeled backhoe. Think about that figure. Does it seem reasonable? How does it compare with past production records? How many hours per day can the backhoe trench? How will traffic control affect daily production? Will daily backfill and cleanup reduce production further? What are the alternatives? Will larger equipment yield a better trenching rate? What's the mobilization cost?

Trenching machines offer high foot-per-hour rates. But the number of crosslines on this job would make a trencher more trouble than it's worth. The only serious contender is a larger tracked backhoe. You know that your crew, using a 1 cubic yard tracked backhoe, will have an average bucket load of 0.75 cubic yards, with a cycle time of 20 seconds (3 cycles per minute):

$$.75 \text{ CY} \times 3 \text{ cycles} = 2.25 \text{ L. CY per minute}$$
$$2.25 \text{ L. CY} \times 60 \text{ minutes} = 135 \text{ L. CY per hour (optimum)}$$
$$135 \text{ L. CY} \times .75 \times .83 \times .6 = 50 \text{ L. CY per hour (realistic)}$$

$$\frac{50 \text{ L. CY}}{.45 \text{ L. CY per LF}} = 111 \text{ LF per hour}$$

Consider these questions. Is this production gain any advantage? Will the size of the project warrant transporting the machine to and from the job?

Job Schedules to Compare Costs
You can now make up job schedules that convert these hourly trenching rates into daily production estimates. For the job schedules, you'll have to plan out the complete job and determine what equipment and crew you'll use for each option.

Figure the complete crew and equipment cost, then add the cost of mobilization. Divide by the linear feet of trench to arrive at a cost per linear foot for each option.

We'll do job schedules for three different possibilities, then analyze the costs and choose the best schedule to fit this project.

Job Schedule 1— Look at Figure 2-7. This is the high production method. Our choice would be influenced by the availability of equipment and distance from the work site. It wouldn't make sense to ship the equipment in from across the state for a two-day job. If the machines happened to be nearby, the mobilization cost would be much less. Of course, the same is true of the other schedules, but not to the same extent. Loading and transporting heavy tracked machines is much more costly than simply driving a wheeled machine to a site.

Because of the small size of this job, mobilization costs are too high for Job Schedule 1. A project of longer duration might make the mobilization cost seem more reasonable and make this method a more attractive choice.

Job Schedule 1 requires a tight schedule. You need near perfect synchronization and no delays. A broken water main, bad weather or equipment breakdown could disrupt the entire schedule. If that happens, the cost per foot would skyrocket. Overrunning the time schedule by just half a day could blow your profit. A highly experienced contractor can avoid some problems, some of the time. If you're not that experienced, or not that lucky, why take the risk?

Job Schedule 2— See Figure 2-8. This schedule offers a good balance of crew and equipment. Mobilization costs are reasonable. Matching the digging hoe and the backfill hoe allows virtually full-time trenching after asphalt has been removed.

Job Schedule 3— See Figure 2-9. This schedule requires a lot of movement by the hoe. The hoe does the excavation, supplies the pipe and bedding and does the backfilling. A variation of this method would be to have a small wheel loader share the supplying and backfilling chores so the 410 has more digging time.

The schedule is a grid chart tracking hourly activities across Day 1 and Day 2. Each activity row is marked with shaded blocks indicating hours of operation.

Job Schedule #1	Day 1: 8 hours	OT	Day 2: 8 hours	OT
"Cat" 950 loader: travel time	▓			
Cutting asphalt	▓░░			
Loading asphalt	░░░			
Pipe bedding & backfill			░░░░░░░	
Truck: haul in pipe bedding	▓░░░		▓░	
Haul out asphalt	░░░			░
Haul out excess dirt				
"Cat" 225: loading & transport to site	▓		░░░░░░	
Trenching	░░░░░░			
JD 510: travel to site	▓		░	
Material layout	░░			
Assist in loading out asphalt				
Supply pipe bedding & compact backfill	░░░░		░░░░░	
Pipe layer: traffic signs & control				
Material layout				
Pipelaying	░░░░░░		░░░░░	
Topman: traffic signs & control			░	
Material layout	░░░░░		░░	
Assist pipe layer			░░░░░	
Work truck	░░░░░░░░		░░░░░░░░	
Supervisor	░░░░░░		░░░░░░	

Mobilization:

"Cat" 950	2 hours @	$50.00 =	$100.00
"Cat" 225	2 hours @	$60.00 =	$120.00
Transport	3 hours @	$60.00 =	$180.00
JD 510	2 hours @	$40.00 =	$ 80.00
			$400.00

Estimated production rate: 7½ hours @ 111 LF per hour = 832 LF per day.
Estimated duration of trenching & pipe installation: 2 days with full crew, not including cleanup and asphalt work.

		Hour	Day
Dump truck and driver	@	$ 35	$ 280
Cat 950 and operator	@	50	400
Cat 225 and operator	@	60	480
JD 510/hoe pack and operator	@	40	320
Supervisor	@	20	160
Pipelayer	@	15	120
Laborer/topman	@	15	120
Work truck and tools	@	4	32
		$239	$1,912
Mobilization	@	$ 400	$ 400
Crew and equipment, 2 days	@	$1,912	$3,824
	Total cost		$4,224

$$\frac{\$4,224}{1,235} = \$3.42 \text{ per LF}$$

Job schedule 1
Figure 2-7

Job Schedule #2	Day 1: 8 hours	OT	Day 2: 8 hours	OT	Day 3: 8 hours	OT
JD 410/cutting wheel: travel time						
Cut asphalt						
Load truck						
Trenching						
Backfill						
410/hoe pack, travel time						
Lift and stack asphalt						
Load truck						
Material layout						
Pipe bedding						
Backfill and compact bedding						
Truck: haul in pipe bedding						
Haul out asphalt						
Haul out excess						
Pipelayer: traffic signs and control						
Material layout						
Pipelaying						
Topman: traffic signs & controls						
Material layout						
Assist pipelayer						
Work truck						

Mobilization:

JD 410	2 hours @	$35.00 =	$ 70.00
JD 410	2 hours @	$35.00 =	$ 70.00
			$140.00

Estimated production rate: 7 hours @ 62 LF per hour @ 434 LF daily.
Estimated duration of trenching & pipe installation: 3 work days, not including cleanup and asphalt work.

		Hour	Day
Dump truck and driver	@	$ 35	$ 280
JD 410	@	35	280
JD 410 and packer	@	37	296
Supervisor	@	20	160
Pipelayer	@	15	120
Laborer/topman	@	15	120
Work truck and tools	@	4	32
		$ 161	$1,288
Mobilization	@	$ 140	$ 140
Crew and equipment, 3½ days	@	$1,288	$4,508
	Total cost		$4,648

$$\frac{\$4,648}{1,235} = \$3.69 \text{ per LF}$$

Job schedule 2
Figure 2-8

Job Schedule #3	Day 1: 8 hours	OT	Day 2: 9 hours	OT	Day 3: 9 hours	OT
410/cutting wheel: travel time						
Cut asphalt						
Lift and stack asphalt						
Load truck						
Trenching						
Pipe bedding						
Backfill						
Truck: haul in pipe bedding						
Haul out asphalt						
Haul out excess						
Pipelayer: traffic signs and control						
Material layout						
Pipelaying						
Topman/handheld compactor: traffic signs & control						
Assist pipelayer						
Compact backfill						
Work truck						
Supervisor						

Mobilization:

JD 410   2 hours @ $35.00 = $70.00

Estimated production rate: 4 hours trenching @ 62 LF per hour = 248 LF daily.

Estimated duration of trenching and pipe installation: 5 days + 1 day preparation work = 6 days, not including cleanup and asphalt work.

(This schedule involves paying two hours overtime, increasing the cost slightly).

		Hour	Day
Dump truck and driver	@	$ 35	$ 280
JD 410	@	35	280
Supervisor	@	20	160
Pipelayer	@	15	120
Topman/hand compaction	@	16	128
Work truck and tools	@	4	32
		$ 125	$1,000
Mobilization	@	$ 70	$ 70
Preparation work, 1 day	@	400	400
Crew and equipment, 5 days	@	$1,012	$5,060
	Total cost		$5,530

$$\frac{\$5,495}{1,235} = \$4.47 \text{ per LF}$$

Job schedule 3
Figure 2-9

Note that this schedule requires overtime work for at least two employees. In this case, the extra cost wouldn't be very significant. The mobilization cost is small. Day one uses only a partial crew costing a total of $400.

But note that Job Schedule 3 is the least efficient and most costly method.

Of course, these schedules are only estimates. They won't be completely accurate. They do, however, let us compare work rates. Each can be changed, of course. The dump truck is included in all three schedules, even though it won't be required all of the time. Renting a truck occasionally or bringing one in from another project when needed would yield some savings. If the truck driver will be on the job every day, he may be able to flag traffic or handle other chores when not actually driving.

The choice is between Schedule 1 and Schedule 2. Based on the available machinery, the projected overall work schedule at the time of the project date, and the availability of the necessary personnel, Job Schedule 2 appears to be the right choice. I would base my estimate on these figures, remembering to add material costs and the cost of removing and replacing the asphalt.

In the last two chapters we've estimated and bid the job and produced a production schedule. At the bid opening, if we were low bidder, we got the work. This is where it gets interesting. We'll start by surveying the site in Chapter 3.

3

READING PLANS AND SURVEYING

The job plans (or *prints* as they're sometimes called) show both the existing features where work will be done and the work you'll have to do. The engineer's responsibility is to provide plans and specifications that explain all job requirements. Your responsibility is to install exactly what the plans and specs show.

You don't have to be an engineer to read and follow plans. Neither do you have to be a surveyor to lay out the job correctly. But you need to understand some basic plan reading and surveying concepts. This chapter is intended to provide the background needed to interpret the plans and lay out the work for nearly any underground utility work.

Let's start by looking at a typical project. Several developers have been active on the west end of your town. The existing sewer system will have to be extended to the proposed tracts. The town council hired an engineering company to study the problem and draft plans for expanding treatment capacity and extending the existing collection system. Acting on the engineer's recommendation, the town council approved the project, arranged funding and authorized the engineers to begin planning the project.

The engineering company sent a survey team into the field to record the actual lay of the land where the sewer line will be extended. This is important information for the design engineer. Without it, he couldn't calculate the falls and slopes in the pipeline. With the surveyor's data in hand, an engineer designed the project.

Next a draftsman drew up plans showing details of the engineer's design. These plans showed the intended work in two different views: a *plan view,* which is a bird's-eye view of the area, and the *profile view,* which shows a horizontal cut-away view of the work.

Reading Plans

Figure 3-1 shows common plan symbols for underground utility work. These are like a draftsman's "shorthand." Different draftsmen may use slightly different symbols, but most sets of plans include a key like this one explaining the symbols used.

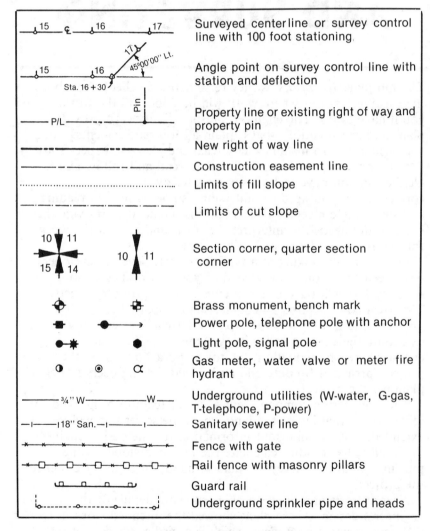

	Surveyed centerline or survey control line with 100 foot stationing
	Angle point on survey control line with station and deflection
	Property line or existing right of way and property pin
	New right of way line
	Construction easement line
	Limits of fill slope
	Limits of cut slope
	Section corner, quarter section corner
	Brass monument, bench mark
	Power pole, telephone pole with anchor
	Light pole, signal pole
	Gas meter, water valve or meter fire hydrant
	Underground utilities (W-water, G-gas, T-telephone, P-power)
	Sanitary sewer line
	Fence with gate
	Rail fence with masonry pillars
	Guard rail
	Underground sprinkler pipe and heads

Common plan symbols
Figure 3-1

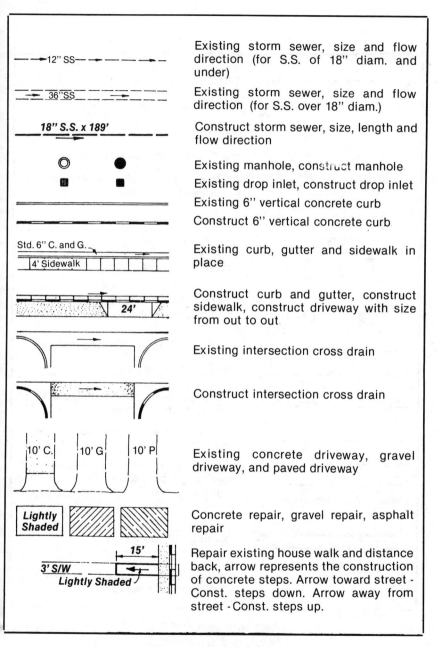

—→12" SS— —→— — — — —→ —	Existing storm sewer, size and flow direction (for S.S. of 18" diam. and under)
—=— 36"SS— — — —=— — — —=— —	Existing storm sewer, size and flow direction (for S.S. over 18" diam.)
18" S.S. x 189' →	Construct storm sewer, size, length and flow direction
◎ ●	Existing manhole, construct manhole
▣ ■	Existing drop inlet, construct drop inlet
	Existing 6" vertical concrete curb
	Construct 6" vertical concrete curb
Std. 6" C. and G. 4' Sidewalk	Existing curb, gutter and sidewalk in place
24'	Construct curb and gutter, construct sidewalk, construct driveway with size from out to out
→	Existing intersection cross drain
	Construct intersection cross drain
10' C. 10' G 10' P	Existing concrete driveway, gravel driveway, and paved driveway
Lightly Shaded	Concrete repair, gravel repair, asphalt repair
15' 3' S/W *Lightly Shaded*	Repair existing house walk and distance back, arrow represents the construction of concrete steps. Arrow toward street - Const. steps down. Arrow away from street - Const. steps up.

Common plan symbols
Figure 3-1 (continued)

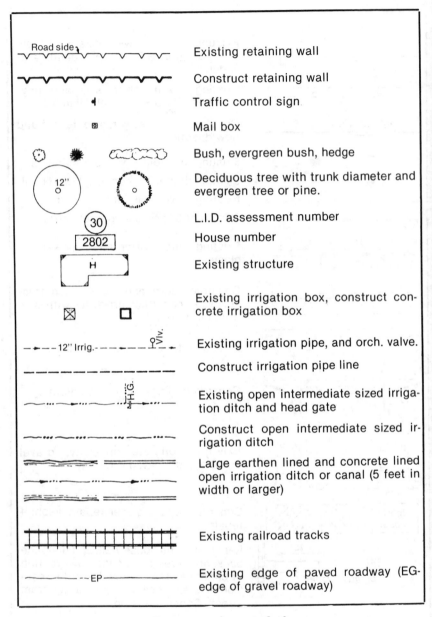

	Existing retaining wall
	Construct retaining wall
	Traffic control sign
	Mail box
	Bush, evergreen bush, hedge
12"	Deciduous tree with trunk diameter and evergreen tree or pine.
30	L.I.D. assessment number
2802	House number
H	Existing structure
	Existing irrigation box, construct concrete irrigation box
12" Irrig.	Existing irrigation pipe, and orch. valve.
	Construct irrigation pipe line
H.G.	Existing open intermediate sized irrigation ditch and head gate
	Construct open intermediate sized irrigation ditch
	Large earthen lined and concrete lined open irrigation ditch or canal (5 feet in width or larger)
	Existing railroad tracks
EP	Existing edge of paved roadway (EG-edge of gravel roadway)

Common plan symbols
Figure 3-1 (continued)

Scale in Utility Plans

The plan drawn could have been drawn full size. But that would make it awkward to handle. Imagine a set of plans 1,000 feet long for a 1,000-foot sewer extension! Instead, nearly all plans are drawn to scale. Many house plans use a scale of 1/4" to the foot, for example. One-quarter inch on the plan is equal to one foot on the actual house. That wouldn't be practical for most utility plans. A 1,000-foot sewer extension would still require a plan sheet over 20 feet long. A scale of 1/16" to the foot would require a plan sheet only 5 feet long. But if the sewer line was 10 feet deep, the scale of 1/16" to the foot would be a vertical height of 10/16", too small to show sufficient detail.

To make your job easier, most plans you use will be drawn in two scales. One scale is used for horizontal distances and another for vertical distances. Usually the horizontal scale is ten times the vertical scale. One inch may represent 50 feet horizontally and only five feet vertically. This lets the engineer show the entire horizontal length of the project in a compact drawing and still include the detail you need for vertical measurements.

Look at Figure 3-2. Elevation numbers run vertically up the side of the print. Both the horizontal and the vertical scales are indicated at the top. Reading two-scale prints takes some practice. Slopes and grades are exaggerated and appear much steeper than they really are.

Surveying

Accurate measurements and precise installation mark professional utility line work. As a professional underground utility line contractor, you'll want to do nothing but precision work. "About right" shouldn't be good enough on your jobs. The engineer won't approve it and you shouldn't tolerate it. If it isn't right on the money, it isn't right. To do accurate work, you'll have to use a surveyor's level correctly.

There are good reasons to follow the plans exactly. Pressure pipelines work best when laid to even, uniform grades. Gravity flow pipelines must be laid very precisely if they are to work as designed. A slope or fall of as little as one foot per 1,000 linear feet is not uncommon in some pipelines. Usually the larger the diameter of the pipe, the smaller the fall required to maintain a water velocity great enough to prevent solids from settling out.

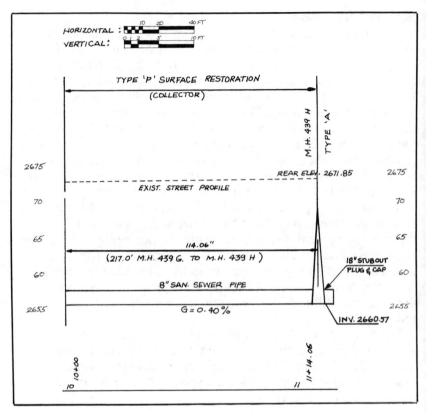

Typical two-scale plan
Figure 3-2

Take the plans seriously. Never assume that the engineer will approve anything different from what the plans show. What you install should give a lifetime of trouble-free service — if installed correctly. Make precision work your company's trademark.

Surveyor's Codes and Abbreviations
Everyone in the underground utility business has to understand the common codes and abbreviations used by surveyors. The way some terms are used may vary slightly from area to area and from company to company. But the basic meaning will be the same no matter where you work.

Elevation, abbreviated El, is the height of a point above sea level. If you know the elevation of two points, you can calculate the difference in elevation between them. If you know the elevation of a certain fixed object, you can calculate the elevation of other points by measuring the vertical distance between the two points.

If the elevation of the outfall of a pipeline is 700.55 feet, and the invert is 706.35 feet, you can calculate the fall in that section of pipeline by subtracting 700.55 feet from 706.35 feet. The fall is 5.80 feet.

It's important for you to understand elevations. Here's another example. Let's say the threshold at your front door is exactly 700 feet above sea level. If you have a 6-foot, 6-inch door opening, the elevation of the top of the opening is 706 feet, 6 inches. In engineering measurements that's El 706.5. We've added the height of the opening to the known elevation of the threshold to find the elevation at the top of the opening.

We've used the threshold as the *benchmark*, the point of known elevation. If your garden path is 3 feet lower than the threshold, you can calculate its elevation: BM (benchmark) El 700.00 minus 3 feet equals El 697.00.

Two other important surveying terms are *cut* and *fill*. A cut (C) is always lower than the reference point. A fill (F) is always higher. The surveyors usually calculate cuts and fills.

It's standard practice to show horizontal distance in multiples of 100 feet. These distances are shown as station numbers. A *station number* (Sta) gives a stake's position in relation to the starting point of the line. The starting point is Sta 0 + 00. The number to the left of the plus sign indicates the number of hundreds of feet. The number to the right of the plus sign indicates feet in units of one. Sta 2 + 50 means that the stake is located 250 feet from the starting point, Sta 0 + 00.

Measuring in Decimal Feet

Engineers use feet and decimals of a foot to measure vertical distances. For example, 2 feet, 6 inches is 2.5 feet in engineer's measure. It's easier to add and subtract feet and decimals of a foot than it is to add and subtract feet, inches and fractions of an inch. For example, it's harder to add 18 feet, $4^{13}/_{16}$ inches to 37 feet, $9^{1}/_{2}$ inches than it is to add 18.41 feet to 37.79 feet.

But don't confuse decimal feet with metric measurements, which are a completely different system of measure. Engineer's

measure uses the good old foot but substitutes decimals of a
foot for inches and fractions of an inch.

In engineer's measure, a foot is divided into tenths and
hundredths. Notice that a hundredth of a foot is very nearly
one-eighth of an inch. Rarely will you see measurements
expressed to the nearest thousandth of a foot (0.001'). Generally
that's more accuracy than is needed in this type of construction.

Engineer's measurements are expressed two ways. For
example, a plan will show an elevation as *El 1520.67*, using the
decimal point to show the portion of a foot. On survey stakes,
the same elevation will be written like this:

$$\text{El } 20^{67}$$

Notice that the first two digits of the elevation have been
omitted and the bottom of the fraction 67/100 isn't shown.

Hubs and Reference Stakes

The engineering company that designed the project will
probably lay out your reference stakes. This survey team should
stay just ahead of your work, providing hubs and reference
stakes with the information necessary to install the pipeline. A
hub is a small stake driven into the ground to identify some
reference point. Near the hub will be a *reference stake* which
identifies the hub's elevation and situation in relation to the
pipeline. Figure 3-3 shows a typical reference stake.

Sometimes reference stakes will show only station numbers,
omitting any other data. In this case, the information usually
given on the stake itself will be on a *cut sheet*. A cut sheet is a
form showing elevations, cuts and horizontal distances to a
fixed point.

Reference stakes along a sewer trench will refer to the depth,
in decimal feet, from the reference hub to the flow line of the
pipe. For example, a reference stake might read:

$$\text{Sta } \quad 2+50$$
$$\text{El } \quad 2500^{47}$$
$$\text{C } \quad 9^{63}$$

This means the flow line of the pipe is 9.63 feet below the
reference hub. The elevation of the flow line of the pipe at this
point is 2500.47 feet minus 9.63 feet, or 2490.84 feet.

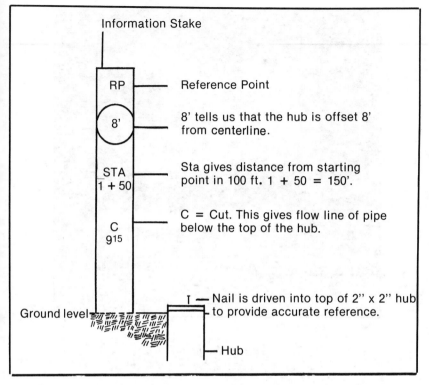

Information Stake

RP ————— Reference Point

8' ————— 8' tells us that the hub is offset 8' from centerline.

STA ————— Sta gives distance from starting
1 + 50 ————— point in 100 ft. 1 + 50 = 150'.

C ————— C = Cut. This gives flow line of pipe
9¹⁵ ————— below the top of the hub.

————— Nail is driven into top of 2" x 2" hub
Ground level ————— to provide accurate reference.

————— Hub

Typical reference stake
Figure 3-3

Surveying Instruments

You're not a surveyor. You don't have to be. But you, and key employees on your crew, should understand a surveyor's job and the instruments used in surveying.

The surveyor's main job is measurement. Horizontal distances can be measured easily with a steel tape measure. Vertical measurements are more difficult. To measure vertical distances (elevations), the surveyor depends on the optical level and the transit. Let's look at each of these.

The *optical level* in Figure 3-4 is a telescope mounted on a tripod. Using the fine adjustment controls on the instrument, you can set it up so your line of sight is exactly level no matter how the scope is rotated.

Looking into the optical level, you'll see cross hairs similar to

Courtesy: Berger Instruments

Automatic level
Figure 3-4

those in a rifle scope. If you sight through the level and line up
the cross hairs on an object several hundred feet away, the
scope's cross hairs will be on exactly the same level as the object
you see. Now pivot the scope so the cross hairs fall on some
other object. Now you know the second object is at the same
elevation as the first.

Because the instrument is mounted on a pivot, you can rotate
the line of sight through a full 360 degrees. This gives you a
level plane which can be used as a reference point over a wide
area and for a distance of several hundred feet.

Let's go back to the example of your threshold for a minute.
If you set up the level in your front doorway and look across at
your garden fence a hundred feet away, the point where the
cross hairs intersect is level with the instrument. If the
instrument is 3 feet above the level of the threshold, the point
you're sighting on the fence is also 3 feet above that level. If
you mark that point on the fence and measure from it to the

top of the fence, you can find the elevation of the top of the fence above your threshold.

An optical level lets you project a horizontal line over considerable distances. It gives you a reference point from which you can calculate the elevation of all points within view *if you know the elevation of the scope.* There lies the key. Once you know the elevation of the optical level, it's a simple matter of addition and subtraction.

Surveyors find the height of the instrument by reading a measuring rod held on a benchmark. They add that reading to the known elevation of the benchmark to determine the height of the instrument, abbreviated *H.I.* If the elevation of the BM is 250 feet, and the rod reading is 2.25 feet, the notation will look like this:

BM	250.00
Rod reading	+ 2.25
H.I.	252.25

The *transit* (Figure 3-5) is a more complicated instrument because it's both an optical level and a device for measuring angles. If you remember any of your high school geometry,

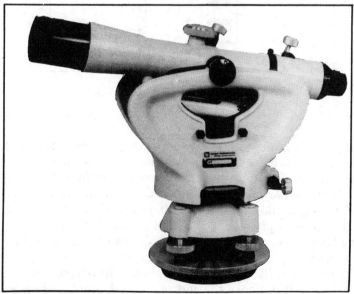

Courtesy: Berger Instruments

Transit
Figure 3-5

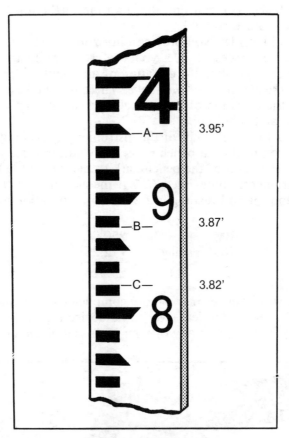

Reading a grade rod
Figure 3-6

you'll recall that if you know two angles of a triangle, you can calculate the third. A surveyor uses the transit to measure both vertical and horizontal distances. We won't go into the method here. That's surveying. You don't need to know surveying to install underground utilities, though you might need to use a transit to align a pipe laser. We'll talk about that later. Nearly all the layout you need to do on the site can be done with an optical level, a grade rod, and a notebook.

A *grade rod* is like a big yardstick that shows feet and decimals of a foot above the end of the rod. Look at Figure 3-6. The long black line with the large 4 near it marks the 4-foot point. Each of the shorter black lines and the white spaces between them measure one hundredth of a foot. The lines with the bottom right corner cut away mark tenths of a foot. Each line with the top right corner cut away signifies a decimal ending in a five.

If the cross hairs of the level intersect at point A, the reading is 3.95 feet. Point A is five segments below the 4-foot mark, and five segments above the 3.9-foot mark. Point B reads 3.87 feet (two segments above the 3.85-foot mark) and Point C reads 3.82 feet. Count the segments between the five mark and the tenth mark to find the rod reading. If you can, practice taking rod readings with someone who's familiar with using a grade rod.

To get an accurate reading, it's essential that the grade rod be absolutely vertical. If it leans away from the vertical, you'll get a false reading. To avoid this, have the person holding the rod slowly lean it toward and away from you while you're taking readings through the level. The lowest rod reading is the accurate one when you use this method.

Using the Optical Level
This is a precision instrument and must be set up correctly. If the line of sight isn't level, the readings taken off the grade rod will be inaccurate. Set up the tripod so it's reasonably level to begin with. Then finish leveling the instrument with the adjusting screws in the instrument base. Centering the bubble in the vial levels the scope. Then tighten the adjusting screw just finger tight when the scope is level.

Now let's put the level to use in a typical job situation. Look at Figure 3-7. The reference stake by the hub is marked C 9.77, showing that you're to make a cut 9.77 feet deep. First, take a reading with the grade rod held on the hub to establish the H.I. (height of instrument). The reading is 5.15 feet. Add this to the cut figure (9.77) to find the H.I. above the required elevation of the finished cut. Here's what you'd write in your notebook·

Rod reading	5.15
C	9.77
H.I.	14.92

Always note where you took the reading and an explanation of the calculations you used to arrive at the revised cut figure.

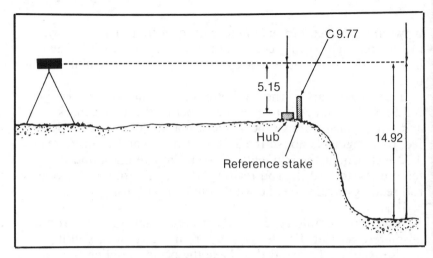

Measuring a cut
Figure 3-7

This reduces the chance of error. A notebook and pencil are as necessary as the level itself for accurate, checkable surveying. You'll use the level most for checking cuts from reference stakes. But sometimes you'll need a level to survey a long distance or a steep slope. If the distance, or the slope, is so great that you can't shoot all points from one setup of the level, you'll have to move the instrument one or more times.

An optical level can only transfer elevations within view. If you can't see some point along the line, set the level up at some new position that has a view of both your reference point and the obscured point. You'll have to establish the elevation of that new position. A quick way to do this is to have the rod holder hold the rod in position while you reposition the level and take a new reading. Then the difference between the two readings is added to or subtracted from the previous H.I., giving the new H.I.

Grades or Slopes in Pipe Laying

Grade is the word usually used to describe the slope of the ground. The term probably comes from the more descriptive word *gradient*. As used in underground utility contracting, it means the desired slope of a pipe, trench or surface.

There are several perfectly good ways to describe the incline of any surface. That can cause some confusion if you're not alert to the problem. I'll cover the common ways of describing slopes, then offer some examples to test your understanding.

Slope in degrees— A circle can be divided into 360 degrees. Each degree is further divided into minutes (60 to each degree) and seconds (60 to each minute). This allows very precise measurement of angles. But, because of the complex mathematics involved, and the precision of the instruments needed to measure slope in degrees, minutes and seconds, this method's seldom used by contractors in the field. Surveyors can measure slope in degrees, but you'll probably never see it on a set of plans.

Slope as a ratio— This is the second method of measuring inclines. It's often used for steeper slopes. The horizontal distance is given first, then the vertical distance. So 3:1 means the vertical drop is 1 foot in each 3 horizontal feet.

Carpenters and plumbers turn it around, expressing the vertical rise first. Plumbers refer to slope in a drain line of 1/4 inch to 1 foot. Carpenters describe roof pitch as 3 in 12, meaning the roof rises 3 inches in each 12 inches measured horizontally. This ratio method is easy to misunderstand and isn't precise enough for utility line work.

Slope as a percent— This is the way many inclines are described in excavation work. The vertical rise is always expressed in units compared to a horizontal distance of 100 units. For example, a 50% slope will rise 50 feet in a horizontal distance of 100 feet. A 1% slope rises 1 foot in each 100 feet measured horizontally. Sewer pipe laid at 0.5% rises 1/2 foot (6 inches) each 100 feet.

Here's an easy way to remember how to figure slope in percent. A 45-degree slope is a 100% slope. If the vertical rise is equal to the horizontal distance, the slope is 100%. Figure 3-8 shows some equivalent slopes expressed in degrees and percents.

To calculate the slope as a percent, divide the vertical rise by the horizontal distance, then multiply the answer by 100. For example, if a hill rises 50 feet in a horizontal distance of 100 feet, here's the calculation:

$$\frac{50}{100} = .5 \times 100 = 50\%$$

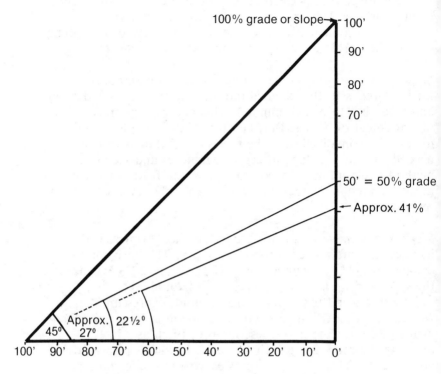

Slopes expressed in percent and degrees
Figure 3-8

Here's a more complicated example. A hill rises 83 feet in a horizontal distance of 227 feet. What's the slope in percent?

$$\frac{83}{227} = .3656 \times 100 = 36.56\%$$

Slope as a decimal— Many underground utility line plans will show pipe grade expressed as a decimal. Percent grades always describe rise in each 100 units. Decimal grades show rise or fall in units as designated by the decimal fraction:

$$.1 = \frac{1}{10}$$

$$.01 \quad = \quad \frac{1}{100}$$

$$.001 \quad = \quad \frac{1}{1,000}$$

$$.0001 \quad = \quad \frac{1}{10,000}$$

Plans usually use four-place decimals to express grade. Let's look at a few examples.

Assume the plans show a pipe grade of 0.0046. That would be a rise or fall of 46 feet in each 10,000 feet of pipe, 4.6 feet in each 1,000 feet, 0.46 foot in each 100 feet, and 0.046 foot in each 10 feet. Reducing the horizontal distance by 10 also reduces the vertical rise by 10. As a percent, that last grade is a slope of 0.46%. So 0.0046 equals 0.46%.

If the plans show a rise of 0.0050, that's a rise of 50 feet per 10,000 feet. Visualize a straight line 10,000 feet long. That's almost two miles. In that distance the pipe rises just 50 feet. A slope that gradual is barely noticeable to the naked eye.

If the slope in percent is 0.25%, the decimal equivalent is 0.0025. That's a rise of 25 feet in 10,000 feet horizontally, or 0.25 foot per hundred feet.

Given the slope or grade in decimals or percent, you may have to calculate how much a pipe or trench needs to rise or fall within a given distance. For example, suppose you're laying a 20-foot length of pipe at a 0.0025 grade. How much must it rise in 20 feet? We can reduce the 25-foot rise per 10,000 feet to 0.25 foot per 100 feet and 0.025 foot per 10 feet. Then double the 0.025 foot for our 20-foot length of pipe. It must rise 0.05 foot in 20 feet. That's five hundredths on the grade rod, approximately 5/8''.

Checking Grade with an Optical Level

Your pipe layers have to understand how pipe grades are measured. Flow rates in gravity pipelines depend on two factors: the inside diameter of the pipe and the pipe grade. The greater the diameter of the pipe, the flatter it can be laid. Engineers understand these principles and plan carefully to use just the right grades. If you don't lay pipe exactly on grade, it won't work right.

High and low spots along a pipeline reduce the carrying capacity. A low spot collects fluids and solids. That increases the friction with material passing through the low spot. A line that isn't on grade may have backfall (reverse flow) which reduces the pipeline's carrying capacity — or it may rise too quickly, causing backflow at another point. Any significant backfall and your engineer will reject the line. The engineer will probably accept minor deviations from a true straight line but will demand precision in grade alignment.

The easiest way to check grade is with an optical level. Sight over from each cut stake provided by the surveyors. Measure down from the line projected by the scope to find the correct elevation at that point. In most cases, it's enough to check grade every 50 feet.

Remember, the cut figure on each stake identifies the flow line of the water — that's the bottom of the pipe's interior. Obviously, you can't hold a grade rod vertical if it's placed on the bottom of the pipe. It's common practice to set the grade rod on top of the pipe. This works fine, as long as you allow for the diameter and wall width of the pipe:

$$H.I. = \text{rod reading on hub} + C - (\text{diameter} + \text{wall of pipe})$$

For 15-inch plastic pipe, the difference between the flow line and the top of the pipe is 1.25 feet. If the cut stake indicates a cut of 12.66 feet and the rod reading was 4.78, the calculation would be:

Rod reading	4.78
Cut	+ 12.66
H.I.	17.44
Pipe	− 1.25
	16.19

The rod reading on top of the pipe opposite the hub would be 16.19.

Laying Pipe with Lasers

Pipe laying requires exact measurement. Pipe lasers offer that kind of precision. Lasers can project a concentrated beam of light in a straight line for several hundred feet. The beam is visible to the naked eye when it shines on a solid surface. Using controls built into the laser, you can direct the beam very precisely along the desired grade.

I've had more arguments over how to set up lasers than over anything else in the pipe laying business. Like any precision instrument, accuracy depends on the skill and knowledge of the user. I'll talk about laser setup a little later. But first, let's see why even this modern instrument isn't infallible.

Light has different characteristics as it moves through air of different temperatures and densities. You've seen heat waves rising off hot asphalt in the September sun. A beam of light actually bends, or refracts, in response to changes in temperature. This problem is worst during extremely hot or cold weather. Pipe that's been in the ground for a while will have about the same temperature all along the line. The temperature of newly buried pipe can vary quite a bit. This causes air currents and heat waves which distort the laser beam as it passes through the pipe. That makes the beam an unreliable reference.

One solution to this distortion is to use a 12-volt fan or blower to blow air through the pipe. This evens out temperature variations and extends the distance at which the laser is reliable.

Check the laser as pipe laying progresses. It's easy to bump or disturb the laser on a busy construction site. If the beam moves, you can bet that there are temperature differences in the pipe. If the beam seems to be enlarged, there's probably water vapor in the pipe or near the laser. I've found the practical maximum range of most lasers to be about 400 feet.

Aligning the Laser

A great deal of time is spent aligning pipe lasers. Much of this time is wasted because the operator doesn't understand the process. Alignment can be very simple. There are several fast and easy ways of doing it.

It would be a snap to set up a pipe laser on a level surface, with good light and lots of working room. It would be easy to aim it at a point several hundred feet away. In practice, however, your laser setups will be in cramped and dark manholes several feet underground. Your target will be an object that isn't in view. This complicates the task considerably.

To project a straight line under these conditions, we need a reference point between the laser and the target. There are two ways to approach this. We could establish a line exactly parallel to the desired line and transfer the parallel line to the laser. A second choice would be to use the offset hubs to record the precise centerline and then transfer it to the laser.

In Figure 3-9 we've measured precisely from the offset hubs to find the centerline. The transit will be set on this centerline before excavation begins.

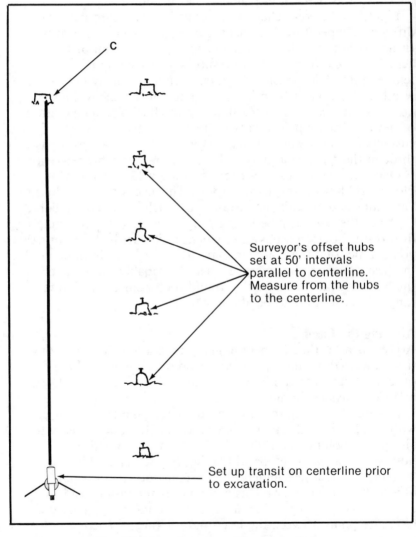

C

Surveyor's offset hubs set at 50' intervals parallel to centerline. Measure from the hubs to the centerline.

Set up transit on centerline prior to excavation.

Establishing the centerline from hubs
Figure 3-9

Suppose we know the starting point and at least one intermediate point along the line from the starting point to the target. We can save time by projecting a straight line along the two points to find the centerline. See Figure 3-10. This takes a precise measurement, however.

To get accurate laser alignment, you need three accurate reference points: the laser, the target (usually a manhole several hundred feet away) and a reference point exactly in between these two points.

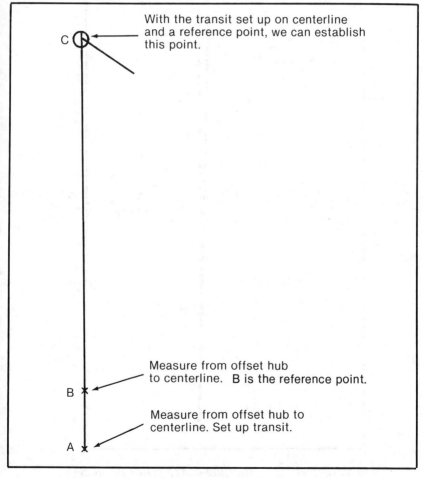

Establishing the centerline with one reference point
Figure 3-10

Figures 3-11, 3-12 and 3-13 illustrate possible alignment errors. In all of these figures, point A represents the laser, point B the intermediate reference point, and point C the target.

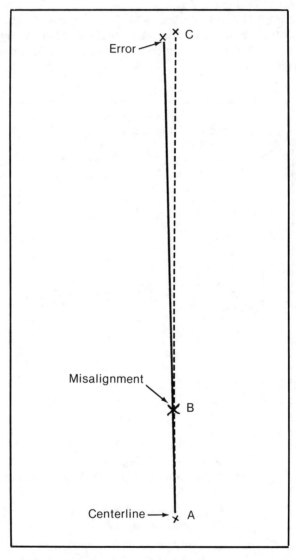

Misalignment of centerline at reference point B
Figure 3-11

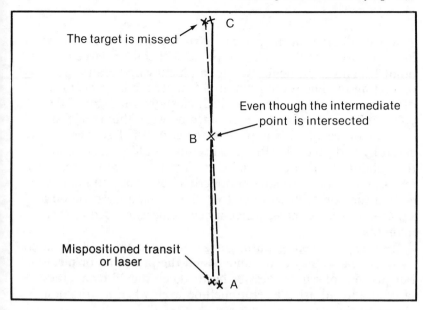

Misalignment of centerline at transit or laser
Figure 3-12

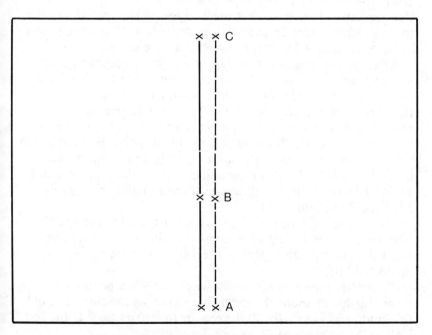

Misalignment of centerline at both points
Figure 3-13

Just a small error in the placement of the instrument or laser, or in the position of the reference point, will cause a large error at point C. This is particularly true in pipe laying because we're projecting the line several hundred feet. If the laser beam is offset but parallel to the centerline, as shown in Figure 3-13, you'll miss the target by the distance between the parallel lines.

You can see why it's important to set up the laser very carefully and precisely. But good setup shouldn't take more than about 15 minutes. Your setup time for the pipe laser is an important and controllable production cost. Save 10 minutes of wasted time on each setup and you have a major saving on a sewer line, for example, where a new setup is needed at every manhole.

Even experienced operators make mistakes when setting lasers — not because they don't understand the principles or methods, but just out of carelessness or haste. Then they'll blame the laser and avoid using a setup method even when it's the best choice in certain circumstances.

Using the Laser to Check Pipe Placement

I've explained how to use a transit to locate the centerline prior to excavation. Once excavation has started, we'll use the laser to be sure the pipe is laid precisely on that centerline.

After part of the trench is dug, use a grade rod and plumb bob to transfer the offset line to the pipe centerline. See Figure 3-14. Place the grade rod on the manhole offset hub and lay it flat across the trench. Suspend the plumb bob over the centerline at point A and place the laser there (Step 1). Then move the grade rod to the next hub and suspend the plumb bob on center point B. Align the laser beam to the plumb bob at point B (Step 2). To check placement, move the grade rod and plumb bob to the next hub and check beam alignment again. See Step 3 in Figure 3-14.

I like to recheck beam alignment by suspending the plumb bob about 50 or 75 feet from the manhole. The further away from the manhole the beam is checked, the more accurate the projected line.

No matter how you set up the laser, there's a potential for error. In this method, the most likely error is in failing to hold the grade rod at exactly 90 degrees to the offset line. If the rod isn't at 90 degrees, you'll get a false reading.

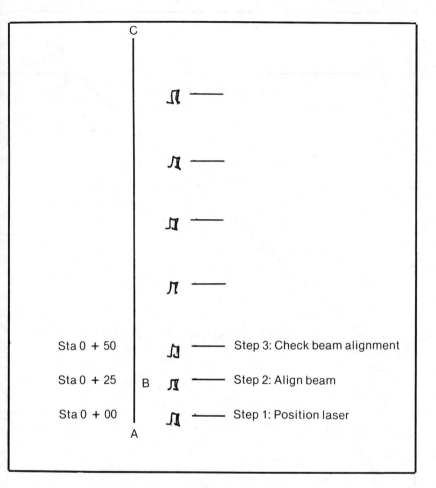

Aligning the laser with grade rod and plumb bob
Figure 3-14

Another reason for inaccuracy is the pressure to set up quickly. If the excavator must be moved off line to set up the laser, production stops while you're measuring. With a large crew and several pieces of equipment, a 15-minute delay can cost you hundreds of dollars. Your best bet is to choose a setup method your crew understands and stick with it.

Using a Transit to Align the Laser
A transit can speed your laser setup *if* the transit is properly aligned to the pipe's centerline. Before excavation, all manholes

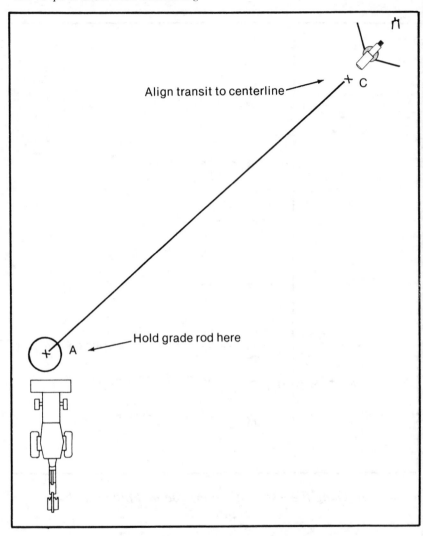

Align transit to centerline

Hold grade rod here

Setting the transit on centerline
Figure 3-15

are given centerline hubs. Set up a transit on the manhole hub upstream from the next manhole to be excavated. Look at Figure 3-15. The transit is set up at the manhole hub, point C, and aligned to the center hub of the manhole to be excavated, point A. When the transit is perfectly aligned to the pipe centerline, lock it on line.

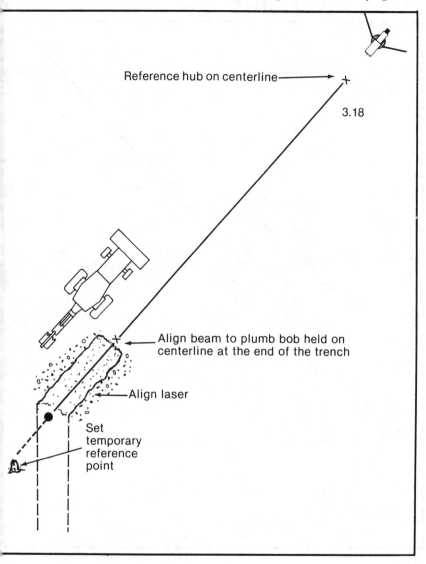

Reference hub on centerline⟶ +

3.18

Align beam to plumb bob held on
centerline at the end of the trench

Align laser

Set
temporary
reference
point

Aligning the laser
Figure 3-16

When the excavator has dug the manhole and a short section
of trench, move the excavator off the centerline. Suspend a
plumb bob over the center of the manhole and set up your laser
as in Figure 3-16. Before resuming trenching, set a temporary
reference point downstream from the laser.

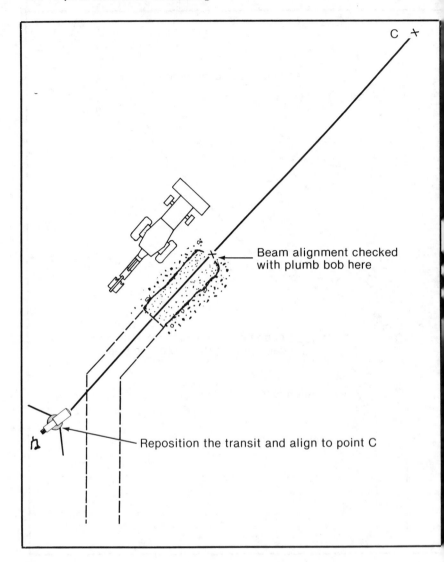

Setting the transit to check the laser
Figure 3-17

Now look at Figure 3-17. You can resume trenching and reset the transit behind the laser. If you move the excavator off the line a second time, you can lock the transit on the centerline. Then you can use the transit to check on the laser as pipe is installed.

The drawback with this method is that the plum bob swings too much in deep trenches and windy weather. It's too hard to find the correct centerline position. Getting the precise position is essential because every small error is magnified by the longer distance between manholes.

The surveyors will usually stake the pipeline at 50 or 100 foot intervals. If this plumb bob method is used, hubs are needed at the manhole and at 25 foot and 50 foot intervals out from the manhole. The surveyors will usually provide the extra hubs if you request them in advance.

A variation of this method is to set up the transit behind the laser at the beginning. The drawback is that it's hard to set up the transit precisely behind the laser without having an established reference point. You have that reference point in the method described previously.

It's the user's responsibility to position the transit in the right spot. Consider using a special transit such as the AGL Special Transit, shown in Figure 3-18. This instrument allows you to align the transit and the laser in one operation.

Courtesy: AGL Corporation

Special transit
Figure 3-18

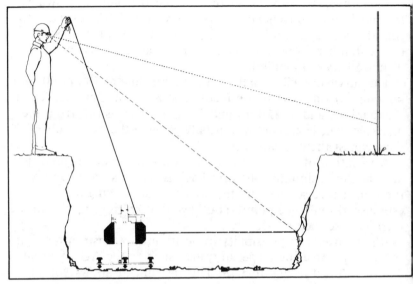

Courtesy: Laser Alignment, Inc.
Setting up the laser by string line
Figure 3-19

Newer lasers are smaller and will fit in small diameter pipe. Some operators try to set the first pipe with a transit and then set the laser in that first length. I don't recommend this. Setting the first length goes slower. Also, any movement of the first pipe during construction will jar the laser out of position.

There's one more alignment method I want to describe briefly. It isn't perfectly accurate, but it's fast and may be accurate enough for large diameter pipe. It's called *string line setup.* Figure 3-19 shows how it's done. First, dial the percent of grade into the laser and establish the proper elevation for the laser light. Then adjust the laser on line. Position a three meter range rod at the center of the next manhole, making sure that it's plumb. Attach a string line in the slot on the handle of the laser, then position the laser in the trench on the centerline of the proposed pipeline. Next, position the string line on the centerline of the pipe, parallel to the range rod, and, using the remote control, adjust the laser light to be on line with the range rod.

Whichever method you use, fast, accurate setup speeds production. But always check both grade and alignment as laying progresses. Check alignment using a plumb bob and rod to establish the centerline from the hubs situated between manholes. Check the grade with a builder's level.

Any of the methods described here, if followed carefully, should give accurate results. But before leaving this subject, I'm going to describe a common problem that has tripped up many contractors. Say you've set up the transit in line with the center of the manhole. But what if the pipe coming into the manhole is cut slightly too long or too short? This changes the position of the center of the manhole. See Figure 3-20. The result is that the established offset hubs or a transit aligned to the proposed center are no longer usable as reference points because they refer to the proposed centerline, not the actual centerline caused by the mispositioned manhole.

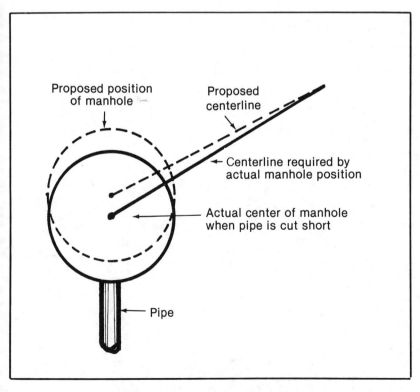

Proposed position of manhole

Proposed centerline

Centerline required by actual manhole position

Actual center of manhole when pipe is cut short

Pipe

Misalignment of manhole center
Figure 3-20

This is probably the most common cause of misalignment, especially when offset hubs are used to transfer the offset line to the centerline. The solution to this problem is to use the surveyor's hubs as a rough guide. Then check it by establishing the position of the laser, establishing the target, and aligning the beam to a point precisely between these points. This means the manhole will be a few hundredths off design center. But that's much better than having the entire section of pipe out of alignment.

4

SITE PREPARATION

Nearly every job requires some site preparation before underground utility work can begin. Preparing the site may be as simple as clearing out some dry grass and weeds that present a fire hazard. Sometimes you'll have to remove dense undergrowth or trees. Slopes may have to be leveled or graded. If you're working in a developed area, you can count on removing pavement and possibly even existing structures. If you need to tear up the pavement before doing any excavation, you'll have to replace it when you're done. Be sure to include the cost of site preparation and restoration in your bid.

In this chapter, we'll examine the site preparation required for underground utility work. We'll start with a look at right-of-way limitations and how to expand your work area. Then we'll examine five common site preparation tasks: clearing grass and brush, clearing trees, preparing the job site, removing pavement and existing structures, and controlling groundwater. The final step in site preparation is material layout.

If you work in Kansas, Nevada or Southern California, there's more information in this chapter on felling trees than you're likely to need. I go into detail on removing trees because it's a major issue for underground utility contractors in the Northwest. If it isn't where you live and work, feel free to skip these paragraphs.

Working Within the Right-of-Way

Most subdivision utilities are placed in the street right-of-way. The right-of-way is often narrow. This means your crew doesn't have much room to work. Working in cramped quarters slows production and requires special procedures. It's always tempting to stray beyond the boundaries of the right-of-way. If it's necessary to increase your work area, *get written permission from the owner of the property before you do it.* This will help protect you against costly lawsuits.

If you can't get permission to work beyond the right-of-way and if the area available is too small for normal operation, your labor and equipment costs will be higher — perhaps much higher. Be sure to take this into account in your bid.

If you're installing a deep sewer in the two lanes of a street or highway, working room is restricted but adequate. If there isn't enough room to pile the spoil beside the trench, it's most commonly loaded out on trucks. To do this you'll need a right-of-way of at least 20 to 25 feet.

Another option is to use a spoil box, shown in Figure 4-1. A spoil box holds the spoil within its walls, until a loader removes it from the box as the excavator places it. If you use a spoil box, a bedding box and a trench box with walls extending above the ground in combination, it's possible to install deep pipe in a right-of-way only 16 to 18 feet wide.

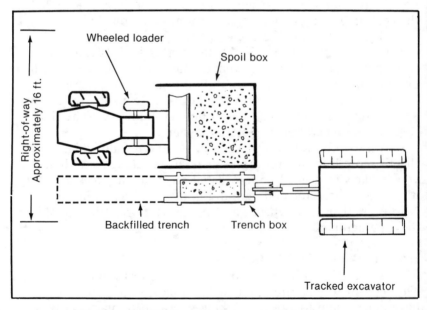

Using a spoil box in a narrow right-of-way
Figure 4-1

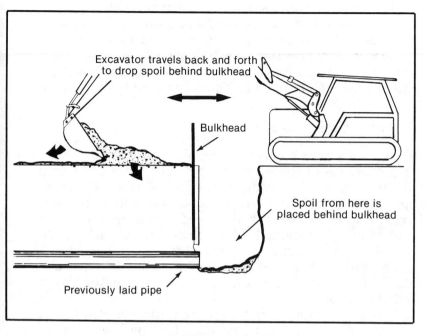

Cut and cover technique
Figure 4-2

On rare occasions you'll be asked to install a line in a 10-foot right-of-way. Here the only option is to cut and cover. Cut and cover, or dig and set, as it is sometimes called, involves placing the spoil from one portion of trench back over the previously laid pipe. Figure 4-2 shows how it works. Before you enter the narrow right-of-way, all spoil is trucked away so there's no backfill over the first section of pipe. Then the spoil from the next section of trench is placed over this pipe. This method requires shoring to minimize the trench width, and an excavator with good horizontal reach. If the trench is relatively shallow, you might be able to use a wheeled backhoe. A tracked excavator is usually more suitable, however, because it can travel back and forth, making spoil placement easier.

It's easier to negotiate a narrow right-of-way if the soil is clay or loam, making the use of hydraulic shoring jacks practical. A bulkhead prevents the loose spoil from falling over the end of the pipe. A sheet of 1-inch plywood will work well with hydraulic jacks, but a sheet of steel will have a longer life.

There's also a third option — a recent and rather ingenious development called the Felco bedding conveyor. It mounts underneath the carriage and between the tracks of the excavator. A hopper stores the bedding material and feeds it onto a conveyor. I haven't seen this tool in action but I've heard good reports about it. You might want to check it out.

Clearing Grass and Brush

Clearing dry grass and brush is simple work. But you know how important it is to remove dry grass if you've ever had a grass fire break out in the middle of your job. One spark from a metal cutting tool can start a blaze when conditions are just right. Provide firebreaks both around the perimeter of the job and around your stockpile of materials. Protect yourself and your investment. Get rid of the fire hazards on your job site.

For brush with a trunk less than 2 inches in diameter, use a dozer. Two or three passes with a heavy disc plow or harrow will do the job.

If you're working in an area where further development is planned, the specifications may not allow you to doze the brush into the ground. But you may be able to doze brush into large piles and burn it. A burning permit may be required. If there's a fire hazard, burning will be out. You may need to rent a chipper and use hand tools to get the site cleared.

For brush that's dense or more than 2 inches in diameter, special equipment may be required. You may want to subcontract the job to a contractor who knows how to maneuver quickly and efficiently in underbrush.

Clearing Trees

If your utility project is in a wooded area, clearing trees may be one of the most expensive items in your bid. Include timber salvage rights in your bid if possible. Salvage rights to commercial grade timber can increase your profit.

Felling and logging trees is a specialty. It can be extremely dangerous if you don't know what you're doing. If there are a lot of trees on the property, consider getting a bid from a specialist. If there are only a few trees, no logging contractor may be willing to do the work at a reasonable cost. It's nice to have someone on your payroll who can handle removal of an occasional tree.

The two pieces of equipment you'll use most often in tree removal are the backhoe and the chain saw. Here's how to use them for safe, efficient tree removal.

Removing Trees with a Backhoe

The best way to remove a tree is to remove the *entire* tree, including the roots. You'll need a midsize wheeled backhoe for this. There are three important steps to follow.

1) Excavate four trenches around the tree, as shown in Figure 4-3A. The trenches should be deep enough to cut through the root system of the tree.

Notice the position of trench D in Figure 4-3A. Dig this trench farther away from the tree than the other three trenches, leaving a platform of dirt between the tree and trench D. The extra dirt will prevent the tree from falling over backward.

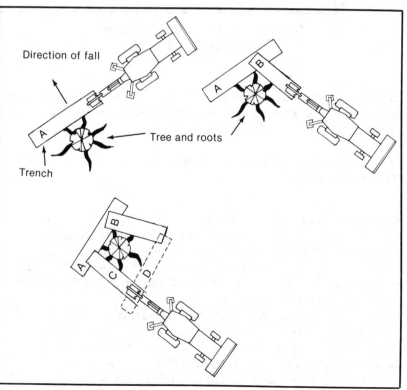

Excavating trenches around a tree
Figure 4-3A

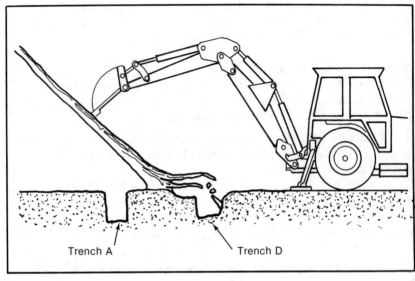

Using hoe dipper to push tree over
Figure 4-3B

2)   Place the hoe dipper against the tree and push the tree over as shown in Figure 4-3B. The higher up the trunk you place the dipper, the more leverage you'll have.

3)   Once the tree is down, use a chain saw to cut the tree into manageable sections.

Removing Trees with a Chain Saw

The modern chain saw is a light, powerful, highly efficient wood-cutting tool that can be used safely. But that doesn't mean that any fool can operate one. A chain saw is like a rifle — harmless by itself or when handled properly, but potentially fatal when handled carelessly.

The two main causes of chain saw accidents are careless handling of the saw and making incorrect cuts that result in unexpected falling timber. Here's how to eliminate chain saw accidents from your site clearance work.

Know and understand *all* of the information in the user's manual before you press the "on" switch. Know how to use the safety devices on the saw and *pay attention to your work every minute the saw is running.* Don't take safety for granted. The saw may be small, but one slip can cost you an arm or a leg.

There's a big difference between sawing up firewood and felling and bucking trees in the woods. Every tree is a complex balancing act. Nearly all trees have a *natural lean*. If you could instantly slice a tree off at its stump, it would fall in the direction of the lean (unless some mechanical force or the wind pushed it some other way).

But "instant slicing" isn't possible with a chain saw. The cutting takes time. And the balance and lean change during the cutting. The size, placement and angle of each cut affects how the tree falls. To control the direction of fall, you have to know how each cut you make is going to affect the natural lean of the tree. Let's look at four types of cuts and how each affects the line of fall. We'll also cover how to cut up the trees once they're on the ground.

Single cut— This is the standard technique for felling small trees up to 6 inches in diameter at breast height (DBH). This method is useful where small trees are growing close together.

Use a single cut to slice the small tree off at the stump. As the saw cuts through the tree, jerk the saw back toward you. The small tree will settle onto the guide bar of the saw, which will pull it off its stump. See Figure 4-4.

Single cut
Figure 4-4

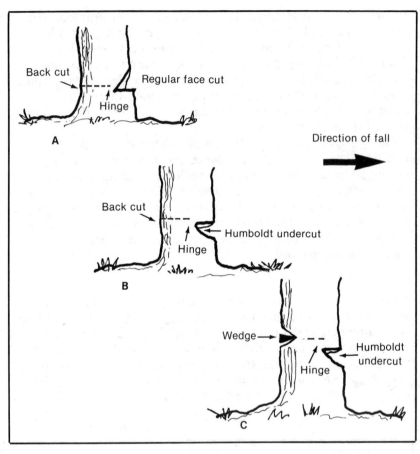

Humboldt undercut
Figure 4-5

Humboldt undercut— This is the conventional method for felling trees larger than 6 inches in diameter at breast height. This method uses the natural lean of the tree. Here are three important steps to follow when using the humboldt undercut:

1) Use a face cut to remove a chock of wood from the stump of the tree. Removing a chock will unbalance the tree.

Look at the two face cuts shown in Figure 4-5. Figure 4-5 A shows a regular face cut. Notice that the regular face cut takes the chock of wood from the trunk of the tree, not the stump that will remain. Now look at the humboldt undercut shown in

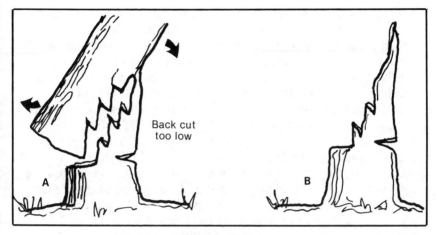

Barber chair
Figure 4-6

Figure 4-5 B. The humboldt undercut takes the chock of wood from the stump rather than from the valuable timber.

2)   Make a back cut on the side of the tree opposite the face cut. The back cut allows the tree to fall in the direction of the face cut. Here are four important things to remember when you make the back cut:

•   The back cut must be on the same horizontal plane as the face cut.

•   The back cut must be slightly higher than the face cut.

•   You *must* leave a hinge of uncut wood running across the stump between the back cut and the face cut. The hinge guides the falling tree, preventing it from slewing sideways or falling in some unpredictable direction. Without the hinge, it may even fall over backward on top of you.

3)   Use a wedge to force the tree past its center of balance. It's shown in Figure 4-5 C.

When using the humboldt or conventional undercut, *you must cut accurately*. If your face cut is too shallow, or your back cut is too low, or the face cut and back cut aren't on the same horizontal plane, the tree may split while falling. This is called a *barber chair*. When a tree barber chairs, the back of the tree swings out and up, pivoting on the top of the split. Figure 4-6 A shows a tree that barber chaired when the back cut was too low. Figure 4-6 B shows why it's called a "barber chair."

Let's look at another example. Assume your back cut is deep enough and the tree begins to fall. But your face cut is too shallow, so the tree can't fall all the way over. The weight of the tree and the leverage of its height are pulling the tree toward the ground. Something has to give. So the tree splits. When the split section of the tree swings out and up, it can hit the feller.

Don't take the barber chair lightly. It's caused many injuries and even fatalities in the logging industry. An experienced tree feller can usually direct the line of fall to any point for a 360 degree circle using the lean, wedges and accurate cutting. But even professionals have been injured by barber chairs. Study the lean carefully. Make accurate cuts. Use the wedges and your equipment correctly. Most of all, be alert when you work.

There's another serious hazard in felling timber, called the widow-maker. A widow-maker is any loose, broken or rotten part of the tree which can fall during felling and hit the feller. Inspect the tree for such loose branches *before* you begin felling.

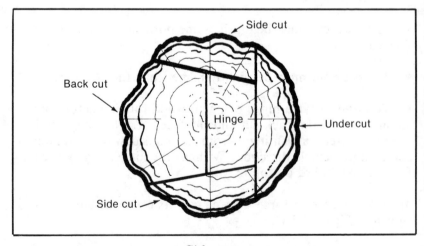

Side cut
Figure 4-7

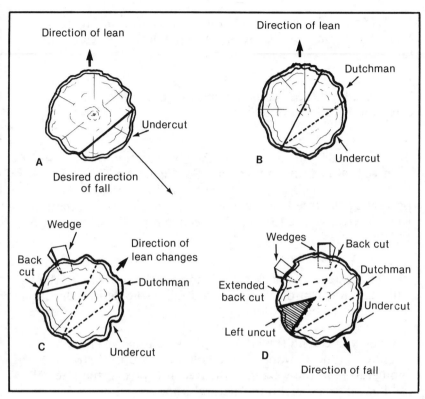

Dutchman cut
Figure 4-8

Side cut— The tree with an extreme natural lean requires a special cutting method. If you use the conventional humboldt method on a heavily leaning tree, it may barber chair even if your cuts are accurate. The extreme lean often causes the tree to fall before the back cut is deep enough. In this case, side cut the tree before you begin your back cut. Figure 4-7 shows how to side cut the heavily leaning tree.

Dutchman cut— Use the dutchman cut when you need to fell a tree *against* its natural lean. Let's look at an example. Assume we're felling a tree at about a 135-degree angle to its natural lean. The tree is 10 inches in diameter. Using wedges isn't practical because they won't lift the base of the trunk enough to compensate for the natural lean. The solution is a dutchman. Refer to Figure 4-8 A through D as we take a step-by-step look at the dutchman.

1) Use an undercut to notch the tree as shown in Figure 4-8 A. Make the cut at a 90-degree angle to the desired direction of fall.

2) Extend the face cut as shown in Figure 4-8 B. This extension is called a dutchman.

3) Use a back cut to cut away the hinge section on the side of the tree that has the natural lean. See Figure 4-8 C. The wood here is under compression, so the cut will try to close up, preventing the tree from falling in the direction of the natural lean. Insert a wedge to keep the tree from slewing sideways. See Figure 4-8 C.

4) Now extend the back cut as shown in Figure 4-8 D. The tree will begin to pivot on its stump and fall in the desired direction. Use wedges to help it along.

When you extend the back cut, be sure to leave enough "holding" (uncut) wood. If the holding wood isn't strong enough to hold the weight of the tree as it pivots, the tree will fall back in the direction of natural lean.

All of these steps happen in just a few seconds. Correct placement of the cuts and wedges is important. If your cuts are slightly off or the wedges aren't properly inserted, the tree may fall in the wrong direction. This is precision work. If your crew doesn't have the skills to perform complicated felling, *don't ask them to do it.* Bring in someone who can do the work safely and efficiently.

Cutting up logs— Once the tree is down and limbs have been trimmed off, the log has to be moved out of the work area. Some trees can be hauled out whole. Others should be cut into pieces first. This is an easy task as long as the log is fully supported along its length. But usually there will be rocks or other obstacles under some part of the log and other parts won't be supported at all. In both cases, part of the wood in the log will be *under compression* and part of the wood will be *under tension.* Here's how to cut a log that doesn't lie flat.

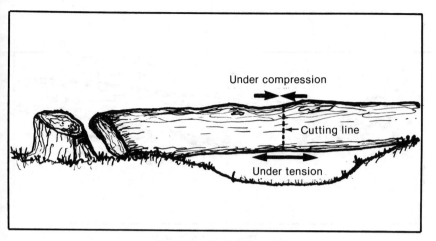

Log spanning a dip
Figure 4-9

When a log spans a dip in the ground, the wood on top of the log is under compression. The wood at the bottom of the log is under tension. See Figure 4-9. If you try to saw this log through from the top, the compression will cause the cut to close up and trap your saw blade while you're cutting. There are three ways to handle this:

• Make two cuts so the saw can't become trapped.

• Use wedges to keep the cut from closing up. Insert the wedges after you cut about one third of the way through the top of the log. Then continue cutting the rest of the way through.

• Cut one third of the way through the log from the top. Then move your saw to the bottom of the log. Finish cutting through from the bottom. This is the simplest and most efficient way to cut a log that is compressed at the top.

When there are rocks or other obstacles under a log, the compressed wood will be on the bottom of the log. The wood under tension will be on top of the log. See Figure 4-10. In this case, your cutting method will be reversed. Cut the first one third of the log through from the bottom. Then move your saw to the top of the log. Finish cutting through from the top.

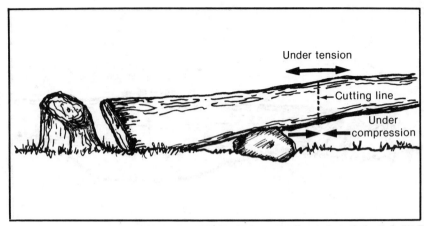

Log resting on obstacle
Figure 4-10

Planning the Work

Site preparation and material layout are important steps in a
well-planned job. In any developed area, your first priority
should be locating the existing underground utilities. These may
include underground power and phone lines, natural gas, storm
drains, irrigation lines, cable TV, water, and sewer lines. You
can usually locate these lines from the building drawings or by
using pipe locating instruments.

But here's the easy way to do it: Most local utility companies
provide a locating service at no cost. But they only want to do it
once. Take the time to record these locations on a set of plans,
or establish an offset or reference point. For example, you can
mark a phone line location on the plans as being 15 feet west of
storm drain inlet.

Spray cans of fluorescent paint are cheap. Use the bright
paint to highlight important structures and existing features. In
many cases, the water meter may be the most accurate reference
to the water service line. If it's painted bright orange, the hoe
operator will be less likely to forget it as the trenching proceeds.

Asphalt Removal

Having marked and referenced the existing utilities, the next
task is asphalt removal. Cutting wheels mounted on a loader,

grader or hoe bucket work well on shallow asphalt. Where the asphalt is more than a couple of inches deep, saw cutting may be required.

Asphalt and concrete slabs are bulky once they've been cut and are ready for loading onto trucks. Allow for a swell factor of 1.7 from in-place volume to loose disturbed volume. In other words, if your project requires the removal of 1,000 bank cubic yards of asphalt or concrete, you'll be handling 1,700 loose cubic yards once it's cut up. Besides being bulky, the asphalt and concrete slabs are hard to handle. A "thumb" fitted to the hoe's dipper helps grasp the slabs and makes loading easier.

If the job involves working on hilly terrain, it's well worth the time and money to build pioneering roads, tracks and stockpile areas.

Material Layout

Material layout may be as simple as stringing out the pipe to be installed. But it can be much more — and it can save you time and money on almost every job. When you're installing water systems, there are often fittings such as valves, crosses and tees that can be assembled and laid out before you move in the main crew and equipment. Two men and a wheeled backhoe will cost you less than $500 a day. If they spend a day or so assembling and laying out materials, it can cut a significant amount of time off the main crew's schedule. That saves money.

There's another advantage to laying out material ahead of time: Material shortages and omissions can be discovered and corrected before they delay a $200-an-hour crew. A $5 gasket can cost you $50 if it delays your crew 15 minutes.

Try to have all needed materials, tools and equipment on hand before trenching begins. And by "on hand" I don't mean just anywhere on the job site. Each material, tool and piece of equipment should be laid out so it's as close as possible to where you'll be using it. If your materials aren't laid out properly, your crew may come to an expensive halt while you run down some missing part.

Sharing tools can also cause production delays. Don't be penny wise and pound foolish: Give each crew a full set of hand tools. And assign someone on each crew the responsibility of keeping track of them. This will help prevent excessive breakage and loss. As in all construction work, a little planning can yield big dividends.

Dealing with Rock

Geology, the study of the earth's surface, is a complex and
fascinating subject worthy of study by anyone who makes a
living moving earth. In this book, we don't have the space to go
into the subject in depth. But there are some broad concepts
which help us understand what rock is, where we're likely to
find it, and how to deal with it. Practically all soils originally
derived from some kind of rock.

Igneous rock is formed from molten magma, the hot liquid
material that lies beneath the earth's crust. Igneous rock that's
relatively young in geological time may be found in an
unweathered, unaltered state. This is sometimes called lava
rock, and it's extremely difficult to work with. It isn't rippable,
and requires a high powder factor to blast it effectively. (The
powder factor is the ratio of explosive to the rock needed to
fragment it.) Igneous rock is very often the parent rock of the
other two rock types you'll encounter.

Sedimentary rocks were formed from sediments (sand, silts
and minerals) that were compressed by water and overlying
layers.

Metamorphic rock may derive from either igneous or
sedimentary rock that was overlaid with other rock and
subjected to intense heat or other pressure which changed the
nature of the parent rock.

Weathering action has broken down all three rock types into
fine particles, creating the many and varied soil types.
Mountains are eroded and washed into the plains, creating clays
and silts often many feet deep. If you live and work solely in
such valleys and plains, you're not likely to encounter rock. If
you venture into hilly or mountainous areas, you'll have to
learn quickly how to deal with rock.

Rippable and Unrippable Rock

For our purposes, we can subdivide rock into two categories:
rippable and *unrippable.* A rippable rock in one that can be
excavated by mechanical means, without the use of explosives.
An unrippable rock can't be moved by conventional earth
moving equipment. It requires blasting.

Generally speaking, sedimentary rock and some metamorphic
rock respond to ripping better than igneous rock, because they
tend to be layered. Rippable rocks include well-weathered rocks
that have fractured and fragmented, most caliches, soft

sedimentary rocks such as chalk, and some sandstone. You can move rippable rock with special-purpose rock buckets or sometimes with special sharp-pointed rock teeth fitted to a conventional hoe bucket.

Obviously, production rates in rock are much slower than in easy-digging clays. In some instances, it makes economic sense to exchange the hoe for a trenching machine. Both wheeled and ladder-type trenchers work well in soft to medium-hard rock and give a good production rate.

Bidding Strategies for Working in Rock

When you bid work in an unfamiliar area, there's always the risk of finding rock. Let's go back to geology for a minute. Mountainous areas were and still are being eroded by water and ice. The ravines and valleys in the mountains become rivers, carrying sediment down toward the oceans. The sediments spread out over the flood plain of the flat areas. Of course, while it's not entirely true that rock is generally found in hilly or mountainous areas, while the flood plains are usually free of rock, it may be taken as a broad generalization.

As you travel to look at prospective contracts, take note of the geography and geology that surround you. A flat-bottom valley usually indicates a flood plain composed of sediment. In this case, you would be more concerned with groundwater and possibly large boulders.

A town situated on a hillside may be sitting a few feet above bedrock. Here are some clues to look for. Do the houses have basements? Is field rock evident?

Bids have been put out with an unclassified excavation clause. "Unclassified ex" means nobody is saying what the soil type is, and you as a contractor have to excavate it, regardless. It's your responsibility to look at the project and assess the conditions. An understanding of how our earth was formed may help avert a very unpleasant surprise.

In many cases, a contract will contain a rock clause, defining rock for the purposes of that job. It may define as rock anything that can't be excavated by a certain size of machine, or boulders above a certain size. Encountering rock that meets this definition is the basis for a change order or extra payment.

Usually, if rock is known to exist, a bid item will be set aside especially for payment for rock excavation. This is a fair way of dealing with the problem. But sometimes engineers don't go to the trouble and expense of exporatory drilling or digging test

Courtesy: ELCO International, Inc.

Plug and feather rock splitter in trench
Figure 4-11

holes. In these cases, you're dealing with an unknown.

If exploratory drilling along a proposed trench line reveals the likelihood of 500 bank cubic yards of rock, you can bid the job on those terms. If it's known that rock exists but no quantity has been established, you may bid based on an estimated quantity of rock. This is a difficult bid, because in some instances you can shoot, remove and haul off hundreds of cubic yards of rock for $50 a bank cubic yard, or even less. Conversely, 200 bank cubic yards can cost you thousands of dollars to blast and remove.

Specialized Equipment for Splitting Rock

Plug and feather rock splitter—Isolated rock humps and large boulders can be broken by the plug and feather rock splitter. This tool works in all types of rock. It can break sections of rock in less than a minute, and is many times faster and more powerful than paving breakers. If large amounts of rock have to be removed, however, blasting would be faster. The splitter is safe when working in a small space like a trench (Figure 4-11), and can be used in urban areas where blasting isn't allowed.

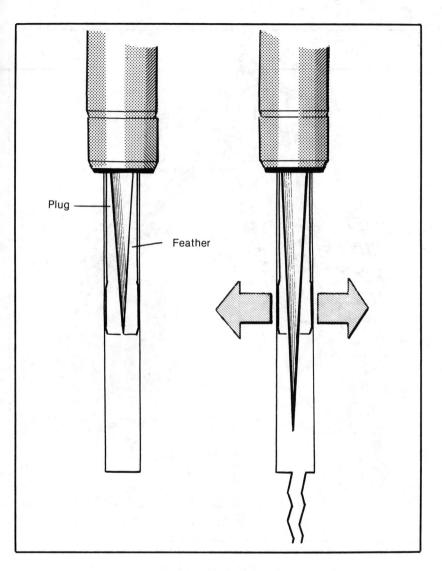

How plug and feather rock splitter works
Figure 4-12

Figure 4-12 shows how a plug and feather splitter works. First a hole is drilled in the rock. Then the plug and feather is driven into the drill hole. As the plug is hydraulically driven between the wedges, the expansion cracks the rock.

Courtesy: Allied Steel & Tractor Products

Hydraulic impact hammer
Figure 4-13

Air-powered and hydraulic impact hammers— Like the plug
and feather splitter, the impact hammer is safe to use in small
areas and where blasting isn't practical. See Figure 4-13. Unlike
the plug and feather splitter, you can use the impact hammer on
all types of rock.

Using Explosives

I'm not going to explain how to be a do-it-yourself blaster. It's dangerous work best left to trained experts. Hire an experienced powder man. A good powder man will make it look easy. A bad one will be a hazard and cost you a fortune in delay time — if he doesn't kill someone. I'll explain only enough so you can recognize the difference.

Where explosives are the only practical way to excavate rock, there are two methods you can use. First, you can deal with each rock mass as you come to it, drilling with hand-held rock drills to set the charges, then covering the charges to prevent rockfly. This is usually the most expensive method because of the unpredictability of when and how much rock will be encountered.

The second and better way is to use an airtrack ahead of the trenching for exploratory drilling. An airtrack is a track-propelled, mechanical drilling rig that's capable of drilling dozens of feet into bedrock. Then you can shoot the rock ahead of the laying crew. This method has the added advantage of using the *overburden* (soil on top of the rock) to hold down the shot.

But here's a word of caution: Even if you have an airtrack available, don't use it to shoot shallow humps of rock up to 2 feet deep. The airtrack drills a large, 3½-inch diameter hole that adds to the risk of the shot going up. Look at Figure 4-14. The shallow, large diameter hole will tend to lift the rock vertically, making it less effective than a small diameter drill hole with tightly packed stemming. *Stemming* is the rock cuttings or sand packed on top of the explosive to reduce the risk of the shot blowing vertically out of the drill hole.

Here's the key to understanding how explosives are used: *The effect of a charge depends on the amount of explosive, the power of the explosive and the resistance it encounters*. There are two common methods of blasting: the single charge and sequential firing.

Single charge— This is the simplest form of blasting. The powder man drills a hole in the rock, places an explosive charge in the hole and sets it off. The shock wave of the explosion travels in the direction of least resistance — up, sideways or, in rare cases, down.

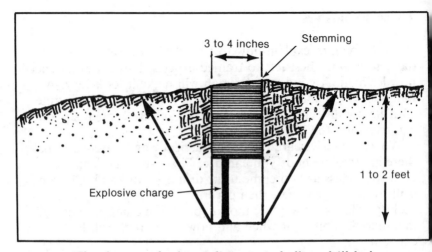

3 to 4 inches

Stemming

Explosive charge

1 to 2 feet

The danger of a large diameter, shallow drill hole
Figure 4-14

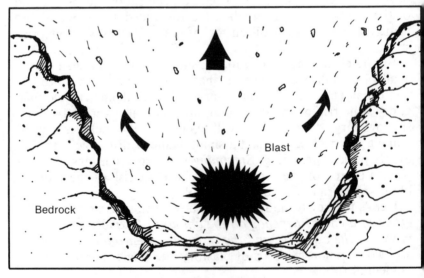

Blast

Bedrock

Single charge
Figure 4-15

If the charge is set in bedrock, the direction of least resistance will be up, as shown in Figure 4-15. If the single charge is set in a small boulder, it should blow the boulder apart.

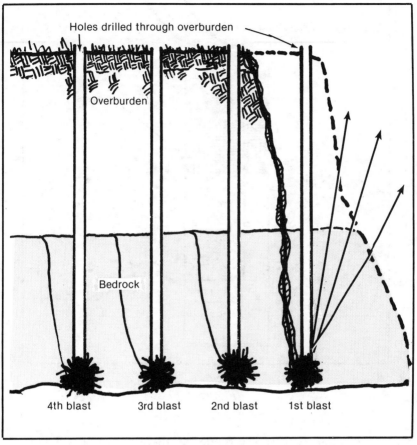

Sequential firing
Figure 4-16

Sequential firing— This is the most efficient and most commonly used method of breaking rock in underground utility line work. The powder man sets up a series of explosive charges in the bedrock. The charges are exploded in sequence. The charge that explodes first creates an open face and a line of least resistance for the charge that comes next. This controls the force and direction of the explosions. See Figure 4-16.

The powder man will usually use electric blasting caps or a detonating cord to provide an initial high-intensity explosion that sets off the main charge.

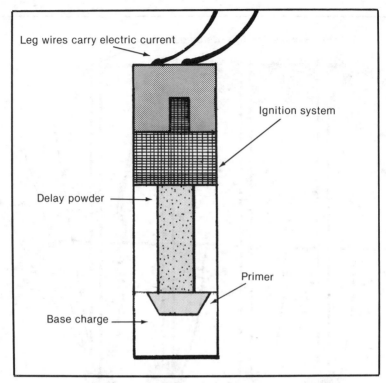

Blasting cap
Figure 4-17

For the charges to fire in sequence, there has to be a delay after the explosion of each charge. Either electric blasting caps or a detonating cord can provide this delay. With det cord, delay is a function of the cord burning rate. Electric blasting caps can have a built-in time delay. The caps are connected to an electric source in series. An electric charge is transmitted simultaneously to all caps, but they explode in sequence according to the time delay in each cap. Figure 4-17 shows an electric blasting cap.

The best way to control the explosion is to use the medium surrounding the charge. An uncovered rock face will produce a lot of flying rock that's thrown out and away from the source of the blast. If your powder man uses the same amount of explosive but sets the charges before the rock is exposed, little or no rock may be thrown out of the excavation. Here's why.

Remember that the effect of the explosion is determined by the force of the blast and the resistance against that force. If the explosion is set off above ground, there's only air to resist the force of the blast. But if the explosion is set off below ground, the surrounding earth resists the force of the blast and absorbs the fast-moving rock. The resistance against the blast is greater than the force of the blast itself. A good powder man knows this and will use it to your advantage. He can detonate a below-ground blast large enough to break up the bedrock while using the overburden to contain and control the blast.

Below-ground blasting can be done two ways. You can drill through the overburden down into the bedrock and set the charges with the overburden still in place. The second way is to remove the overburden, set the charges and then cover up the charges with spoil. For extra protection, use a blasting mat.

Be careful when covering a charge with spoil. The charges are linked together by tiny wires. The powder man will use a galvanometer to check these wires to be sure the circuit isn't damaged when you're backfilling. If any wires are broken, he can't set off the charge. The circuit will have to be repaired before detonation. Tracing wires through loose dirt set on top of live, primed charges is dangerous work.

A good blast will break up the rock without damaging existing utility lines or structures. A bad blast may cause damage and leave rock that can't be excavated and must be reshot. If the blaster doesn't drill the holes deep enough or doesn't use enough powder, the shot may leave a hump of solid rock near or above pipe grade. If there's a weak point in the bedrock below the explosion, the shot may even go down instead of up. If you have to dig out or redrill and reshoot a hump, there will be about a two-hour delay for your laying crew.

As we've just seen, blasting extensive rock is a lot more complex than just moving dirt around. And rock work probably isn't finished when the blasting is done. The specifications usually demand that you haul the rock off the site after you've broken it up. Then you have to haul in fill to replace the rock. Be sure to allow for these additional costs in your bid.

Working on Steep Terrain
If the terrain is too steep, you won't be able to level the work area. You'll have to use equipment suited to working on slopes. The light equipment you use for working on level ground may

not be safe on a slope. You may need cables and heavy equipment to anchor the lighter equipment in place. Tracked equipment usually works better on slopes than wheeled equipment.

Working in Groundwater

Surface water from rain and irrigation seeps down into the soil, forming underground pools. The level of the underground water is known as the *water table*. Because of varying soil conditions, the water table doesn't occur at a constant depth below ground. It can range from a few feet to hundreds of feet. In many areas, the water table varies drastically in depth, varying from area to area, season to season and even street to street.

When the level of underground water is near the surface, it may be caused by an impermeable layer that lies near the surface. The water can't flow through the layer, so it remains near the surface of the ground.

Groundwater causes several problems for the utility contractor. It creates unstable soil that's dangerous to work in and impossible to use as backfill. It can flood your trenches, washing out all the work you've done so you can't finish the project.

High groundwater should be detailed in the specifications. You should know about it before bidding the job. This is one area of site preparation where you may want to be *overprepared*. Three groundwater conditions that you should know how to handle are: saturated soil, seeping water, and flooded trenches.

Saturated Soil

When silt and clay are saturated with water, they're impossible to compact. On jobs where compaction is important, you'll probably have to haul silt and clay spoil off the site and haul in replacement fill. The saturated spoil won't be suitable for fill as long as it's wet. It may be acceptable as fill once it's dried out. If you're working on a large job, there may be time for the spoil to dry out. Otherwise you'll have to haul in replacement fill.

In pastures and other areas where compaction isn't a problem, you can use sloppy spoil as backfill. The fill will slowly assume the same characteristics as the surrounding soil.

When granular soils, such as sand and gravel, are saturated, they drain quickly and will be reasonably stable even when damp. Some dampness may even aid compaction of this type of soil.

Seeping Water

Where underground utilities must be installed below the water table, special techniques are needed. Even shallow trenches 5 feet deep are difficult to work with when the water table is high. The problem is that the water saturates the soil, making it unstable.

Water seeping into an excavation acts as a lubricant, allowing large slabs of dirt to slip out of place, causing dangerous cave-ins. When water seeps into granular soils, the soil begins to run or flow. This is known as *flowing sand* and it's extremely dangerous. The drier top layer is quickly undermined by the layer moving underneath it. Collapse is likely. To protect your crew and your work in areas where there's seeping water, use shoring. Drain rock that's at least 2 inches in diameter will provide a firm foundation for your pipelines.

In some cases, very fast trenching, with pipe laying and backfill immediately following the trenching hoe, will handle the problem in shallow trenches. You'll also need to use porous pipe bedding to allow space for the water under the pipe. The idea is to dig, lay and backfill at a speed that outpaces the infiltration.

Flooded Trenches

With deeper trenches in a high water table, you're facing a more complex problem. The trench walls are unstable, usually requiring shoring or coffins — and in addition you may go far enough into the water table to cause the water to "boil up" into the bottom of the excavation.

Have you ever tried to force an air-filled balloon into a tub of water? If so, you know how strong the pressure of the water is. In the same way, the water surrounding the trench exerts pressure on the water in the bottom of the excavation. So in a trench 10 feet below the water table, the water enters the trench under pressure. It's not seeping, it's rushing into the trench.

Here we have a nasty problem. The upthrust may be strong enough to "blow up" the trench bottom, so you don't have the stable foundation necessary for pipe laying. This bubbling, boiling up action can make it virtually impossible to excavate. When a bucket full of soil is removed, it's replaced almost immediately by waterborne soil.

I've had the dubious privilege of working in these conditions, both as a pipe layer and equipment operator. So I know it *is* possible to get pipe in the ground under these conditions. The first and most important lesson is that there's no such thing as too much pumping capacity. If your pump won't handle the water, you can't dig the trench.

Large diesel-powered trash pumps are the best choice, or where semi-permanent pumps can be used, electric-powered submersibles. I don't like to use electric pumps where the pipe laying is taking place; a severed power cord or problem in the pump could get someone electrocuted.

The ideal spot for a semi-permanent electric pump is in the manhole downstream. Plug the outfall line and simply let the pump clear out the water flowing down the pipe to it. Given enough pump capacity at this point, we only have to deal with the water below pipe grade. But if you're laying in a permeable soil like sand and gravel several feet below the water table, that water below pipe grade can be thousands of gallons per minute.

Use the newly-laid pipe to channel the groundwater— If possible, allow the groundwater to flow down the newly-laid pipe to the manhole. Then excavate ahead of the pipe. Figure 4-18 shows the sequence of work. A weld mesh screen is placed against the end of the last pipe, buried in a course of clean rock (usually 1/2-inch to 3/4-inch diameter). The screen and rock allow the water to flow to the pump in the previous manhole.

Using a sump— This water will be muddy and full of silt. It can quickly block the pump intake hoses and may need to be run through a baffle tank before it's discharged. To prevent blocking the pump's intake hose, use a sump. The sump filters the water before it arrives at the pump intake. Use scrap corrugated steel or plastic pipe. Cut a series of slots in it with the pipe saw, place the pipe in the hole and surround it with pea gravel. Now place the intake hose inside it. Pretty easy? Not always. In saturated soils, getting just a few feet deep is a

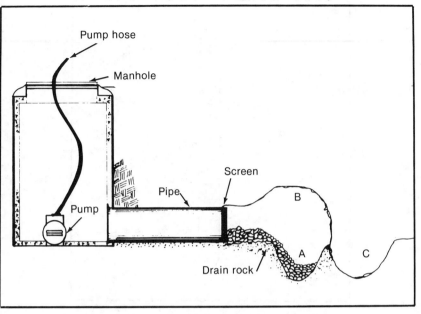

Sequence of excavation in high groundwater
Figure 4-18

challenge requiring cooperation between the pipe crew and the
hoe operator. You'll probably have to place the sump
immediately behind the hoe bucket as it is drawn forward.

It will also take lots of drain rock to stabilize the flowing
sand or earth so the sump can be installed. Once the sump is
installed, you can dewater the trench and lay the pipe.
Then the trench is gradually excavated in short deep sections,
about 2 feet deep and a few feet long. The first section to be
excavated is labeled A in Figure 4-18. As the hoe pulls the
sloppy earth away, the rock piled behind it slips down to fill the
hole. Then pile more rock in the area marked B. Excavate the
area marked C, allowing the rock in B to slide into the
excavation. Repeat the sequence until there's sufficient room to
lay a pipe.

If you choose to lay pipe using this technique, remember that
production will be very slow. You'll also use large quantities of
rock.

Running the groundwater out through laid pipe usually leaves
sludge deposits. The screen and drain rock can keep some silt

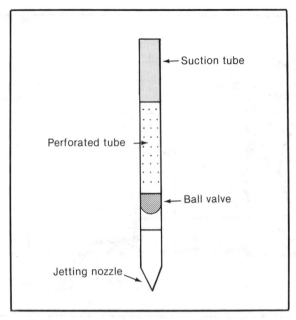

A wellpoint
Figure 4-19

from entering the pipe. If you do find sludge deposits in the pipe, they should be cleaned out. In small-diameter pipe, use manual or mechanical flushing to clean out the sludge. In large-diameter pipe, a worker may need to get inside the line to clean it out. This is dangerous work. If a pump fails somewhere up the line, a heavy flow of water may flood the line, putting the worker in extreme danger. Make sure you have enough dependable equipment to handle any emergency.

Here's another useful tip. Cover the laid pipe with a fairly nonporous fill. This will help prevent water following the pipeline to the area where you're laying new pipe.

Wellpoint Dewatering
If you have a lot of pipe to install in bad groundwater conditions, consider dewatering ahead of pipe laying. Wellpoint systems work very well in sandy soils. They usually have problems in clay or hard overburden where it's difficult to get them down through the surface. They simply won't penetrate sand and gravel with large boulders. Figure 4-19 shows how a

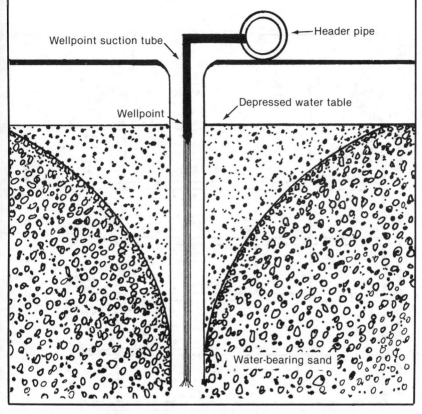

A single wellpoint lowering the water table
Figure 4-20

wellpoint uses water pumped through the jetting nozzle to penetrate the earth. Figure 4-20 shows a single wellpoint at work. Figure 4-21 illustrates a wellpoint system working to depress the water table.

If wellpointing is needed, solicit bids from subcontractors. Wellpoint subs jet well holes at 10 or 15 foot intervals along each side of the proposed trench and to a depth below the level of the trench bottom. A pump is used to suck water out of the wellpoints as long as necessary. When pipe is laid and the trench is refilled, the wellpoint contractor removes his wellpoints. Obviously, this can be a very expensive procedure. But at least it's a cost you can identify ahead of time.

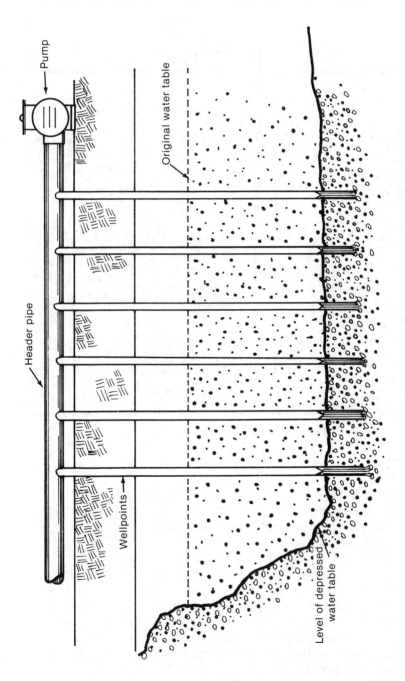

A wellpoint system depressing the water table
Figure 4-21

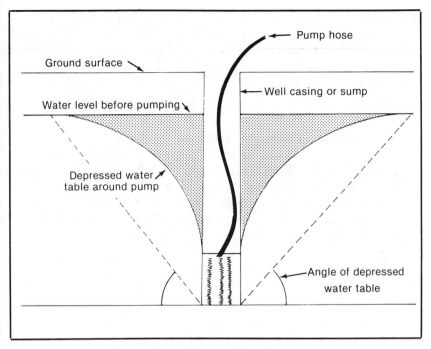

Cone of depression
Figure 4-22

If wellpoints won't penetrate, you need to talk to your local well drillers. One technique you can try is to drive well casings several feet below pipe grade, place a slotted PVC sewer pipe inside the casing, surround the PVC pipe with pea gravel, and pull the casing. You can have workable dewatering sumps at intervals on each side of your trench line.

The cone of depression— Groundwater and its behavior is a branch of engineering called hydrology. It's very complex and beyond the scope of this book — but there's one important concept you need to know. It's called the *cone of depression,* illustrated in Figure 4-22. The angle of the cone of depression depends on the permeability of the soil, the quantity of water drawn from the bottom of the cone, and the speed with which it's drawn. Fast pumping gives a steep angle of depression, while slower, sustained pumping flattens that angle. You're unlikely to achieve an angle of depression of more than 60 degrees over a short time period. That means you need to space the sumps close together, or from 6 to 10 feet below pipe grade, to adequately dewater a trench line. The greater the depth of your dewatering sumps, the greater the area that can be dewatered — *if* you have sufficient pump capacity and discharge facilities.

Production Rates in High Groundwater

Anytime you're working in high groundwater, count on lower production rates. Pump failures can slow the work. Your earth-moving equipment will be working at a reduced capacity. Average bucket loads won't exceed the bucket's struck capacity. The practical capacity of your trucks will be less. Be sure to allow for this in your bid.

5

OPERATING
A BACKHOE

With this chapter we begin covering the key pieces of equipment every utility line contractor uses. We'll begin with the most common piece of heavy equipment on utility line jobs, the wheeled backhoe.

A backhoe can be the most versatile piece of construction equipment on your underground utility line jobs. It may also be the most challenging to operate. It's versatile because it handles so many jobs so well. It's challenging because every task requires variations of the different skills and techniques of backhoe operation.

Backhoe productivity depends on the skill of the operator. Most excavation equipment takes a minute or more to complete one work cycle. The loader backhoe can complete a digging cycle in less than 10 seconds — if the operator can make a complex series of precise moves without wasted time or motion. It's important that *each* work cycle be efficient. Ten wasted seconds may not seem worth worrying about, but multiply that ten seconds by several thousand times each working day, and you're talking a lot of money.

In this chapter we'll discuss how to get the most out of your backhoe. We'll look at the parts of a backhoe and how they work. We'll discuss the precision skills your operator needs and suggest ways to develop those skills. I'll recommend a good way to calculate backhoe production rates. Then we'll look at the two backhoes you'll be using, the wheeled backhoe and the tracked backhoe. Finally, I'll describe what I feel are the keys to operating these two machines both efficiently and profitably.

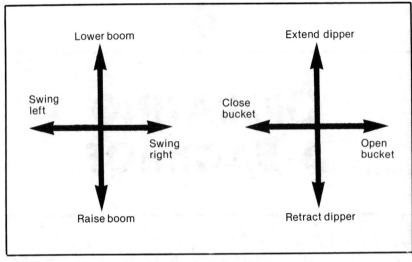

Common hoe controls
Figure 5-1

How the Backhoe Works

The hoe has four main parts: the boom, the dipper (or stick), the bucket and the swing mechanism. The operator has eight different ways he can move these components, using the control levers on the operator's platform. Figure 5-1 shows the most common wheeled backhoe controls. Figure 5-2 shows other control panel layout options found on wheeled backhoes.

But efficient backhoe operation is more complex than just making eight separate maneuvers. Here's how the control maneuvers combine in a typical *digging cycle* for a wheeled backhoe.

Part of digging cycle	Control maneuver
Load bucket	Retract dipper
	Lift boom
	Adjust bucket angle
Lift load	Lift boom
	Adjust dipper
	Adjust bucket angle

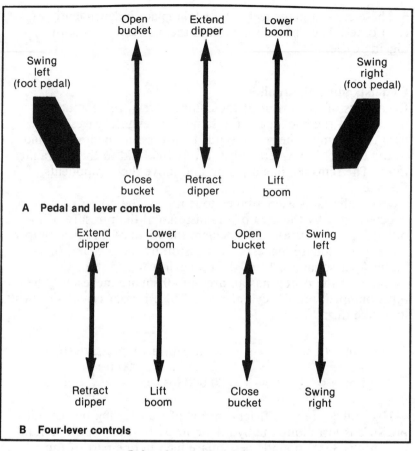

Other hoe control layouts
Figure 5-2

Part of digging cycle	Control maneuver
Swing and prepare to dump	Swing
	Adjust dipper
	Adjust bucket angle
Dump bucket load	Dump bucket
	Extend dipper
	Lower boom
Return to dig	Return swing
	Adjust dipper
	Adjust bucket angle

These control maneuvers don't take place independently of each other. The operator combines them to give a smooth digging cycle.

Understanding Hydraulics

Regardless of the layout of the control levers, the operating principle of the backhoe is the same. The engine drives a hydraulic pump. The pump sends high-pressure hydraulic fluid to the control box where valves direct the fluid to the hydraulic rams. The rams act like a piston and move the components of the hoe.

Hydraulic rams exert tremendous force. The power of a ram is determined by the area of the piston and the pressure of the oil against that piston. To calculate the force of a ram, multiply the area (in square inches) of the piston by the pressure (in pounds per square inch) of the oil against that piston. For example, if the piston has an area of 10 square inches and the oil is pumped into the cylinder at 2,000 psi, your calculation will look like this:

$$\text{Force (lbs.)} = \text{Area (sq. in.) x pressure (psi)}$$
$$\text{Force} = \text{10 sq. in. x 2,000 psi}$$
$$\text{Force} = \text{20,000 lbs.}$$

Hydraulics are an efficient means of transferring power. Little pressure is lost along the hydraulic lines. If your hydraulic fluid pressure is 3,000 pounds per square inch in one part of the system, the hydraulic pressure will be very nearly 3,000 pounds per square inch at every other part of the system. There's hardly any loss of force due to friction in a hydraulic system. Pressures in backhoe hydraulic systems usually range from 2,000 pounds per square inch to 3,000 pounds per square inch.

Regulating the Response Time

The control valves give you precise control of the oil flowing into each cylinder. By controlling the rate of flow, you control the response time of the equipment without affecting the power of the ram. If you reduce the flow of oil into a cylinder, the ram will move more slowly *but with the same force.* When you open the control valve all the way, the flow of oil increases, the piston moves faster and extends the ram faster.

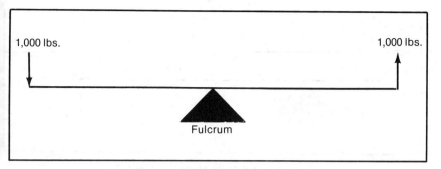

Equal leverage: 1/1 ratio
Figure 5-3

Always use the control valves to regulate response time. *Do not reduce engine speed to slow your response time.* The backhoe engine is designed to run at a specific speed. It's less efficient if you run it below the design speed during excavation operations. And the backhoe design assumes a certain hydraulic pump output at a given engine rpm. Reducing the rpm can cause the hoe to "misbehave" because the pump output is insufficient.

Using Maximum Power
The ram is most efficient near the middle of its travel. Here's why.

All of a backhoe's rams operate at a *leverage disadvantage.* This means that a small movement of the ram results in a larger movement at the end of the lever. Figure 5-3 shows what equal leverage looks like. If the leverage is equal, and there's 1,000 pounds of force applied to one end of the bar, the same amount of force is transmitted to the other end of the bar. Notice the position of the fulcrum. When the fulcrum is in the middle of the bar, the leverage is equal.

In Figure 5-4, the fulcrum is no longer in the middle of the bar. Now when you apply 1,000 pounds of force to one end of the bar, only 333 pounds of it is transmitted to the other end. This is a leverage disadvantage of 1/3. The leverage disadvantage of a backhoe dipper usually ranges between 1/3 and 1/6, depending on the length of the dipper. The leverage disadvantage of a dipper changes as the dipper moves. The backhoe is most powerful when the ram has the least leverage

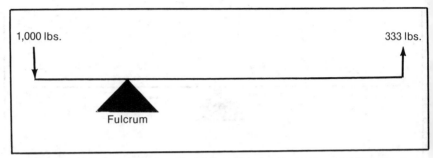

Leverage disadvantage: 1/3 ratio
Figure 5-4

disadvantage. This is usually when the ram is near the middle of its travel. As the ram extends, the leverage disadvantage changes slightly. The ram is most favorably positioned near 90 degrees to the boom. When the ram is at the end of its travel, the leverage disadvantage is at its greatest. Figure 5-5 shows the ram at its greatest leverage disadvantage, when fully retracted and fully extended.

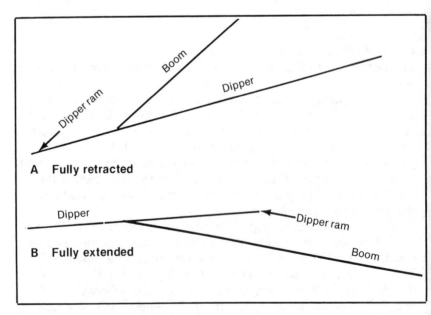

Changing leverage disadvantage
Figure 5-5

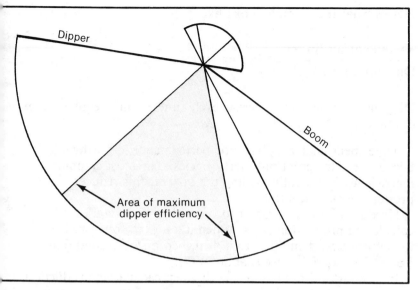

Maximum dipper efficiency
Figure 5-6

There are times when you may need to use the maximum reach of your backhoe. But a skilled operator will keep most work within the maximum efficiency area shown in Figure 5-6. When work is outside the easy reach of your equipment, it's time to reposition the hoe. It's more productive to reposition your equipment than to struggle with work at the end of the reach.

Calculating Backhoe Production Rates

In production trenching, the backhoe excavates one *set* at a time. A set is the length of trench that a hoe can excavate from one position. A set can be broken down further into digging cycles.

Here's how to calculate the time required to excavate a set. First, divide the volume of dirt in the set by the dirt moved per cycle. This gives you the number of digging cycles required per set. Then multiply the number of digging cycles per set by the average cycle time to get the total time it takes to excavate a set.

Here's what the formula looks like.

$$\frac{\text{Volume of dirt in set}}{\text{Dirt moved per cycle}} = \text{digging cycles per set}$$

Digging cycles per set x average cycle time = time required to excavate a set

Time spent analyzing backhoe performance is worthwhile because it helps point out inefficiencies. The most common inefficiency is not fully loading the bucket; this reduces production significantly.

Here's an important point to remember when you're calculating production rates. When it's moved, dirt swells from its natural *bank* condition to a disturbed or *loose* condition. Loam, for example, swells about 30%.

It's important to allow for swell when calculating production rates. Here's how to do it. Let's excavate a 2-foot wide by 4-foot deep trench in loamy soil. To find the number of bank cubic feet of dirt per linear foot of trench, multiply the depth of the trench by the width of the trench. The trench contains 8 bank cubic feet of dirt per linear foot.

Now multiply the number of bank cubic feet by 1.3 (the factor for loam) to get the number of loose cubic feet. Next, multiply the number of loose cubic feet by the number of feet in a set. This gives you the total *loose* volume of dirt in the set. Your calculation will look like this:

Trench depth x trench width = bank cubic feet per linear foot of trench
Bank cubic feet x soil factor = loose cubic feet per linear foot of trench
Loose cubic feet x number of feet in set = total loose volume of dirt in set
4 ft. deep x 2 ft. wide = 8 bank cu. ft. per linear ft.
8 bank cu. ft. x 1.3 (factor for loam) = 10.4 loose cu. ft.
10.4 loose cu. ft. x 8 ft. set = 83.2 loose cu. ft. in set

Before we compute the final production rate, let's look at bucket capacity. The heaped capacity is usually about 130 percent of the struck capacity. For example, a 2-foot wide hoe bucket has a struck capacity of 5 to 6 loose cubic feet. So the heaped capacity would be 130 percent of this amount, or about 6.5 to 8 loose cubic feet.

Assume your operator gets heaped capacity bucket loads of 8 loose cubic feet per bucket. His cycle time is fast, averaging 12 seconds. And it takes him 15 seconds to move his machine into position for the next set. The total time required to excavate a set will look like this:

$$\frac{83.2 \text{ loose cu. ft. in set}}{8 \text{ loose cu. ft. per cycle}} = 10.4 \text{ digging cycles per set}$$

10.4 digging cycles x 12 seconds per cycle = 124.8 seconds
124.8 seconds + 15 seconds to move machine = 139.8 seconds

$$\frac{139.8 \text{ seconds}}{60 \text{ seconds per minute}} = 2.33 \text{ minutes to excavate one set and move machine to next position}$$

$$\frac{60 \text{ minutes per hour}}{2.33 \text{ minutes per set}} = 25.75 \text{ sets per hour}$$

25.75 sets per hour x 8 linear ft. per set = 206 linear ft. of trench per hour

If you alter the cycle time or bucket load in the formula, you'll see quite a dramatic increase or decrease in production.

Such apparently small increases are the difference between average and above-average operator production.

Operator Ability

Anyone can operate a backhoe. It's just a matter of moving the levers. Given a little time, anyone could find a way to scratch the ground and dig a hole. But that's not what a professional operator gets paid for. A good operator has control so precise that he could open a soda can with his bucket. He's aware that good control, good positioning of his machine and maximizing the amount of material per cycle leads to efficient, productive work. A perfectly executed cycle, if the hoe bucket is filled to less than capacity, is still a wasteful cycle.

The operator's control levers open valves that allow hydraulic fluid to flow to the rams that activate the hoe. The wider he opens a valve, the greater the flow to the ram, and the faster the ram will operate. Reducing the flow slows the speed of the ram. A skilled operator will always use the control valves — *not the engine speed* — to increase or decrease the speed of the rams.

Highly skilled operators can use a hoe almost as an extension of their body. They develop a "feel" for obstructions in the soil or changes in ground condition. Of course, it's impossible to feel a steel cross-line before you get to it. But a skilled operator will feel the slightest pause in the progress of the hoe on first contract. When the bucket strikes a solid object, the tractor will pull toward the obstruction.

When the hoe hits a fragile cross-line, it usually happens too quickly for even the most skilled operator to prevent damage. A wheeled backhoe hoe will slow only imperceptibly before cutting through a 50-pair phone cable. But if the operator knows the general location of a cross-line, he may be able to detect a difference in soil resistance as the line is reached. In certain soils, soil resistance is usually greater in undisturbed earth than in an existing trench line. The key is paying attention to the work — looking for trench lines or soft spots in otherwise firm soil. It results in fewer surprises and less damage to cross-lines.

It's essential that the operator be able to *hold grade* — keep the trench at the correct depth. Nearly all trenching work requires that the excavator maintain a specific grade. For example, water pipe installation specs may demand an even grade and a minimum cover over the entire length of the pipeline. Sewer service lines have to flow toward the main line and the trench bottom must have a precise grade. The operator has to deliver the grade called for in the specs.

The operator has to keep the bottom of the trench even along the length of the trench. Overdigging is a waste of both time and material. And leaving humps or a rough, uneven trench bottom makes hand grading necessary. A good operator will minimize the handwork and make the hoe do as much of the work as possible. A highly skilled operator can keep the trench bottom even without constant directions from the grademan.

A good operator can "eyeball" an even grade. He does this by noting the position of the dipper as he draws it toward the machine. He notes where the dipper is in relation to the top of the trench. See Figure 5-7. When the bucket is at the correct

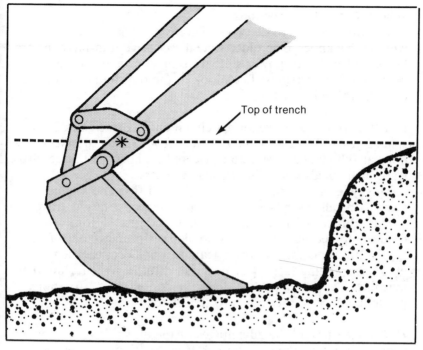

"Eyeballing" an even grade
Figure 5-7

depth, the operator eyeballs a spot on the dipper and makes sure that spot stays parallel to ground level as he continues to dig. This keeps the trench bottom parallel to ground level.

The Wheeled Backhoe

The advantage of a wheeled backhoe is that it's easy to maneuver and can double as a loader. If any piece of equipment deserves the title of all-around workhorse, this is it, the tractor-loader-backhoe. Its specialty is digging trenches less than about 8 feet deep.

First we'll look at the three types of transmissions fitted to the wheeled backhoe. Then we'll look at some key pointers for positioning the wheeled hoe. Finally, I'll describe five specialized tasks wheeled backhoes usually handle in underground utility work.

Wheeled Hoe Transmissions

There are currently three types of transmissions fitted to
wheeled backhoes. The older style direct drive transmissions are
being replaced with power shift transmissions and torque
converter transmissions. Let's look at the pros and cons of each
type of transmission.

Direct drive has a conventional clutch and six to eight closely
spaced gears. The travel speed ranges from 1.5 to 20 miles per
hour at full engine speed. John Deere backhoes with direct drive
transmissions also have full-power reversers.

This type of transmission has low-speed first and second
gears. This allows full-power hydraulic response at low travel
speeds. But constant double clutching and gear shifting is
required to get maximum front loader performance out of this
type of transmission. Shifting gears is usually easier when the
motor is spinning fast, at or near full throttle. When using these
machines, set the hand throttle at the desired rpm (rotations
per minute). Save the foot throttle for road travel.

Power shift has all the advantages of the direct drive
transmission plus the ability to shift gears at any engine speed
without using the clutch.

Torque converter   Most later model backhoes are fitted with
torque converter 4-speed transmissions. The torque converter is
a fluid coupler. There's no direct contact between the engine
flywheel and the transmission. Engine torque is transmitted
through the transmission fluid.

There are two advantages to the torque converter
transmission. It reduces wheel spin. And it can reduce loader
cycle times when the machine is doing shuttle work such as
truck loading.

In my opinion, the drawback to this type of transmission is
that first gear may be too fast for certain types of work. There
are times when you need low travel gear, no greater than 1 to
2 mph (miles per hour). For example, when the backhoe is stuck
in deep mud or trying to climb a steep or slippery slope, you
may need to use both the dipper and the transmission to push
the hoe forward. If the machine is fitted with a 3- to 4-mph first
gear, the transmission either stalls, or minimum travel speed is
simply too fast to do any good.

Selecting the Right Size Hoe
A backhoe is most efficient when working within 70 percent of its maximum digging depth. Here's an example. If you're excavating a trench that's 10 feet deep, you'll need a hoe with a maximum digging depth of about 14 feet. To find the maximum digging depth of the hoe required, just divide the trench depth by 70 percent.

$$\frac{\text{Trench depth}}{70\%} = \text{maximum digging depth of hoe required}$$

$$\frac{10 \text{ ft. deep trench}}{.70} = 14.3 \text{ ft. maximum digging depth required}$$

This isn't to say that the backhoe can't be used to dig deeper trenches. It's just more efficient in this range.

A hoe with a maximum digging depth of 14 feet has a horizontal reach of 16 to 17 feet, measured from the swing post of the hoe. This size backhoe can excavate a 6- to 9-foot section of trench in one position.

If the soil is loose sand and gravel, the backhoe may not have enough reach to place the spoil far enough away from the trench. An extending dipper will increase both the digging depth and reach for spoil placement. It's usually more efficient to use a larger machine, however, if you have a lot of deep excavation.

The typical wheeled backhoe has a digging depth of 14 feet. Larger models such as the Deere 710B and J.I. Case 780C have digging depths over 17 feet. It may be more economical to rent a larger wheeled hoe for deeper trenches, or look at the possibility of a tracked excavator. Usually, you're better off paying the extra for the most suitable equipment rather than suffering slow production rates.

Now that you've selected the right size backhoe for the job, let's talk about how to position it for maximum efficiency.

Positioning the Hoe
When digging around or under obstacles, it's important to have the hoe in a good position. Change the position of the hoe several times rather than work from a bad position. This is especially true when excavating under a sidewalk or working around cross-lines.

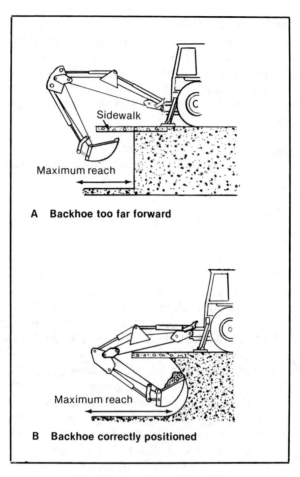

A   Backhoe too far forward

B   Backhoe correctly positioned

Digging under a sidewalk
Figure 5-8

If your backhoe is correctly positioned, you can tunnel completely under a 4-foot wide sidewalk. Figure 5-8 shows how to do it. In Figure 5-8 A, the backhoe is too far forward and can't reach all the way under the sidewalk. In Figure 5-8 B, you can see that moving the machine backward a few feet lets the hoe reach completely under the sidewalk.

Quick, accurate positioning will increase your trenching production rate. There are three ways to move the hoe into the right position. You can (1) use the hoe to push the tractor forward, (2) use the transmission to move the hoe, or (3) use the hoe in combination with the transmission to push the backhoe ahead.

Using the hoe to move the tractor forward— Raise the front bucket and lift the rear wheels and stabilizers off the ground. The tractor will then roll forward on its front wheels. Another way is to leave the rear wheels on the ground and lift the stabilizers just enough so that the tractor can roll on all four wheels. Fully retracting the stabilizers is only necessary when you need to lift the stabilizers anyway to clear an obstruction.

Using the transmission to move the hoe— Most backhoes have shuttle levers mounted near the steering column. You can engage the transmission with the operator facing toward the rear. But it may be hard to control the machine in this position. It's best to stay in the driving position when you're using the transmission to move the hoe.

Backhoes fitted with torque converter transmissions require throttling down the engine rpm to achieve a slow travel speed. Backhoes with direct drive or the Deere powershift will creep ahead at 1 to 2 mph at full throttle.

Using the hoe to push the backhoe ahead— Figure 5-9 shows how to do this. The disadvantage is that it may disturb the bottom of the trench, especially in soft ground. Consider setting the dipper off to the side of the trench, as shown in Figure 5-10. This can be a difficult maneuver. The tractor will tend to swing off the centerline, as shown in Figure 5-10 A. Correct this by steering the wheels to the left to counteract the tendency of the tractor's front to move right, and by using the swing control of the hoe to offset the push away from the centerline. See Figure 5-10 B.

Working Around Cross-Lines

When working in developed areas, you're going to have cross-lines. Some are just a nuisance, others are dangerous. But all of them are expensive to repair. Using a grademan to locate cross-lines *before* you hit them prevents unnecessary damage.

If there are many cross-lines, locating all of them can be a full-time job. You'll need to hire someone to do just that. If there are only a few cross-lines and the pipe laying work isn't too heavy, the pipe layer can double as grademan.

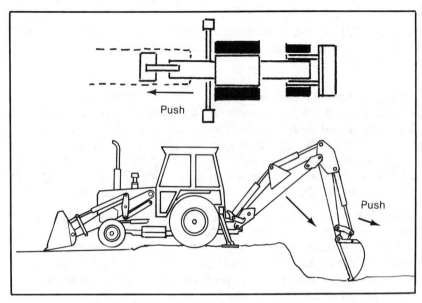

Using the hoe to push the backhoe ahead
Figure 5-9

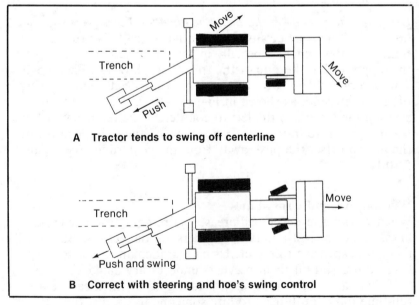

Setting dipper off to side of trench
Figure 5-10

Most cross-lines have been laid in a trench. By carefully watching the walls of the ditch, the grademan can often spot the old ditch line. Look for darker spots of topsoil mixed in with lighter-colored subsoil. The soil in the old ditch will usually be easier digging than undisturbed soil.

Some cables and flexible pipe are ploughed into the soil without any trenching. There won't be many clues to their whereabouts. But most utility companies will locate and mark their lines. *Always* take advantage of this service. And mark the location of known cross-lines on your blueprints. You can use these to refresh your memory if one of the utility company's markers gets wiped out.

If a utility company marker does disappear, have them come mark it again. An additional safeguard is to paint an offset reference mark in a position where it's unlikely to be covered up or destroyed. Some utility companies have a "guaranteed" safe area, usually one foot beyond their paint mark. If they mismark the location and the cross-line is in fact outside this area, they'll pay for any cross-line damage. If mislocation is common, report it to the construction department of the utility company. Be sure everyone understands that the fault is theirs and not yours.

Modern pipe detectors are good for locating metal pipe from above ground. But nonmetallic water and sewer lines can't be located with metal detectors. Fortunately, these are the cheapest pipes to repair.

Finding the cross-line is only the first step. The next is to excavate around it. The best technique is to excavate underneath the line, as shown in Figure 5-11. Here are the four important steps:

1) Remove the dirt from the area *behind* the cross-line, as shown in Figure 5-11 A. The operator should keep the pipe in view the entire time he's working near it. Treat the line as though it were made of glass. Even pressing dirt against an old pipe may make it snap.

2) Remove the dirt that has been piled up against the back of the pipe, as shown in Figure 5-11 B. When removing the dirt, don't pull it in toward the pipe. Shove the dirt back and away from the pipe.

3) Move the backhoe ahead. Quickly excavate *in front of* the cross-line. See Figure 5-11 C. Make sure the hoe is in the right position before you do this.

Excavating around cross-lines
Figure 5-11

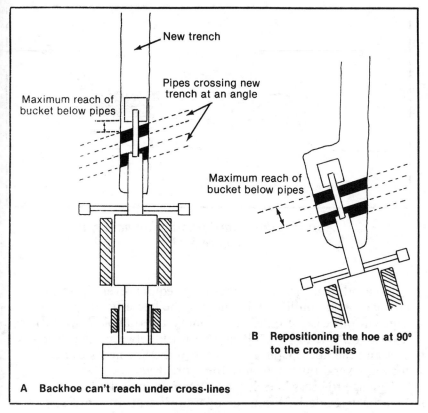

Cross-lines crossing new trench at an angle
Figure 5-12

4)   Move the backhoe into a position where it can easily reach under the cross-line. Remove the loosened dirt under the line, as in Figure 5-11 D. Always flick the soil down and away from the cross-line. You may be able to excavate a tunnel under the cross-line for your new line. Don't remove any more dirt than necessary. Remember what I said before: It's better to reposition the backhoe rather than dig from an unfavorable position.

When the cross-lines cross the new ditch at an angle, the backhoe may not be able to reach far enough to excavate underneath them. See Figure 5-12 A. Solve this problem by moving the backhoe into the position shown in Figure 5-12 B before you excavate. Careful positioning of the hoe is crucial.

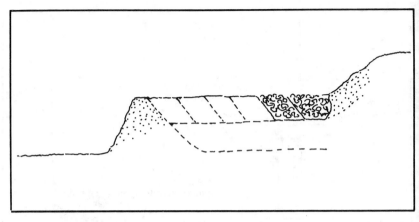

Applying downward pressure on hard soil
Figure 5-13

Excavating around cross-lines may require several position changes. And you'll only be moving the hoe a few feet each time. In this case, use the hoe to move the tractor forward. It's a waste of time to switch back and forth between the driver's seat and the hoe operator's seat when the hoe needs to be moved several times but only for short distances.

Damage to cross-lines is more common in rocky soil. The hoe hits rocks regularly. From the operator's seat, it's hard to distinguish a rock from a cross-line when you first hit it. The machine tenses up or pauses. You may be encountering existing utility lines. Or you may just be running into natural rock.

Hard Digging

In normal digging conditions, you load the bucket by pulling it toward the machine. In hard ground you'll have to use other techniques. Two methods for handling hard ground are shown in Figures 5-13 and 5-14.

One way to break hard material is to force the bucket downward into the soil, breaking off a section of material with each thrust. See Figure 5-13. Begin in the section of trench nearest the machine. Once that material is broken up, use the conventional pickup technique to remove the soil.

The second method is to remove the material in arcs, slicing off a few inches of material with each pass. See Figure 5-14. This pulls the new material toward already loosened earth.

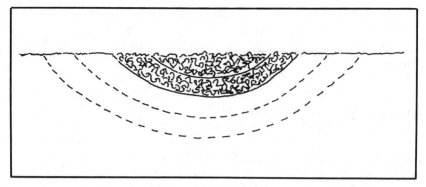

Removing hard soil in arcs
Figure 5-14

Whether your operator uses downward thrusts or a slicing motion depends on what works best. The technique is to pull the hard material towards broken soil. It's more effective than simply tugging at the hard-to-move earth.

No backhoe can dig in solid rock. But your operator can be expected to handle softer materials, such as caliche, cemented sand and weathered rocks. Some operators use what I'll call the "drop and smash" method in material like this. They use the dipper as a ram, crushing the material into fragments. I won't claim that this doesn't work. It does. But it also tends to cause cracks and breaks in the boom and dipper. Backhoes aren't designed to take this kind of abuse. Use it in an emergency, but understand that drop and smash greatly shortens equipment life.

Sewer System Work
Your operator will usually dig trenches from the low end to the high end. It's usually easier to stay on grade when the trench slopes up toward the machine. But sometimes you'll have to dig a falling grade. For example, if you dig a sewer line from the property line to the main line, grade will slope down toward the machine. Let's look at an example. We'll install a 4-inch PVC sewer service. Assume it's 35 feet from the main to the property where the service is required. A well-kept lawn prohibits taking the backhoe onto the property. At the property line, the pipe needs to be 5 feet deep for the specified minimum grade of 0.002 foot. This is 0.02 foot (2/100 foot) vertical rise per 10 feet of length.

To determine how deep to install the riser, multiply the distance from the main to the property line (35 feet) by the vertical rise required (0.02 foot) to get a depth of 0.7 foot. This means that the riser must be installed at least 0.7 foot *deeper than the pipe at the property line*. Allowing for 0.5 foot of bedding gravel, the trench needs to be dug 5.5 feet deep at the property line starting point and fall to 7 feet by the time it gets to the riser pipe (assuming the surrounding ground is level).

Always use a grademan for this type of trenching. He'll indicate when you reach the 5.5-foot depth and will monitor the depth as you proceed toward the main line. If the ground is fairly level, he can use the ground as a reference point. If not, use a small hand sighting level.

We know the trench needs to be 1.5 feet deeper at the riser pipe. Let's divide the trench up into thirds and check our progress along the way. The first checkpoint is at about 12 feet out from the property line. The depth at this point should be 0.5 foot deeper than at the property line, or 6 feet deep. The second checkpoint is at 24 feet out from the property line. Here the trench should be 6.5 feet deep. The third checkpoint is the riser pipe, and here the trench should be 7 feet deep. You can see how important the grademan is in this type of work.

Water System Work

Whether you're installing a system in a new subdivision or repairing an existing system in a developed area, you'll likely use the wheeled backhoe to install water services. You'll need a trench from the main to the property line and an enlargement around the meter box. The water main will need a tap or saddle, requiring working room around the main.

In a new subdivision, your excavation work may simply involve digging a trench from the main to the property line. Trespassing won't be a problem when you're working on undeveloped lots. But if you're doing replacement work in an area with existing homes, your work area may be limited by the surrounding private property. If you need to trespass, get permission *in writing* from the owner of the property. But you'll still probably have to repair any damage. It's better to stay off private property.

You may also have to contend with an existing meter box and service. These will probably be in use (under pressure) and will feed off the existing main line. Enlarging the work area around

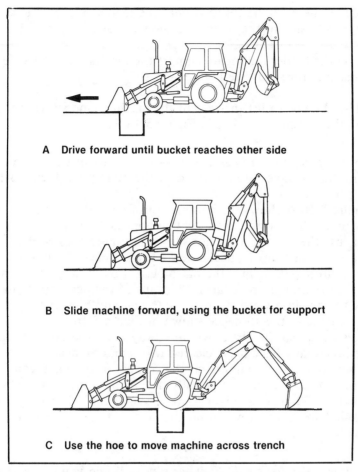

Moving the hoe across an open trench
Figure 5-15

an existing meter box can be tricky. The existing service pipe
may be laid in a direction you wouldn't expect. Take your time
excavating this area and use a grademan.

If you work back from the property to the main, it may
involve specialized tasks: moving a hoe across an open trench,
excavating toward an open trench, punching holes for pipeline,
and excavating the work area for a saddle or tap.

Moving the hoe across an open trench— Figure 5-15 shows the
three steps required to move a backhoe across an open trench.

1) Drive the machine forward until the loader bucket reaches the opposite side of the trench, as shown in A in Figure 5-15.

2) Use the loader bucket to support the machine as you slide the machine forward. See Figure 5-15 B.

3) Use the hoe to lift the rear of the machine and shove it across the trench, as shown in Figure 5-15 C.

Excavating toward an open trench— As you excavate toward the new main, take care not to cover the main pipe. Figure 5-16 shows how to do this.

Figure 5-16 A shows the hazard from falling dirt at the junction of the two trenches. As you approach the junction point, use the digging sequence shown in Figure 5-16 B to excavate the last dirt from the new trench. Be careful not to exert a direct pull toward the machine. If you excavate this area using conventional procedures, the walls of the new trench are likely to cave in toward the existing trench and bury the pipe. If you excavate in the sequence shown in Figure 5-16 B, the final strokes will push the dirt down and away from the existing pipe. Cave-ins usually have to be cleared with hand excavation.

Where two trenches intersect, the corners of the bank can become unstable. This is especially true if the trench intersects at less than 90 degrees. Figure 5-16 C shows the extremely unstable bank formed when the intersecting trench comes in at an acute angle.

Punching holes for small diameter pipe— When you come to a cross-line, don't bother tunneling all the way under it. Instead, stop trenching at the cross-line and begin the trench again on the other side of the line. Then punch a hole through the bank of soil under the line and thread the pipe through the hole. Most water service pipe is flexible and can be threaded under cross-lines.

A good tool for punching holes through a bank of soil is a 2-inch diameter steel bar about 5 or 6 feet long with a point on one end and a plate welded on the other end. This can be shoved under obstructions in soft soil. It leaves a neat 2-inch diameter hole when it's withdrawn. A midsize backhoe can shove or pull a 2-inch bar through several feet of soft soil without any problem.

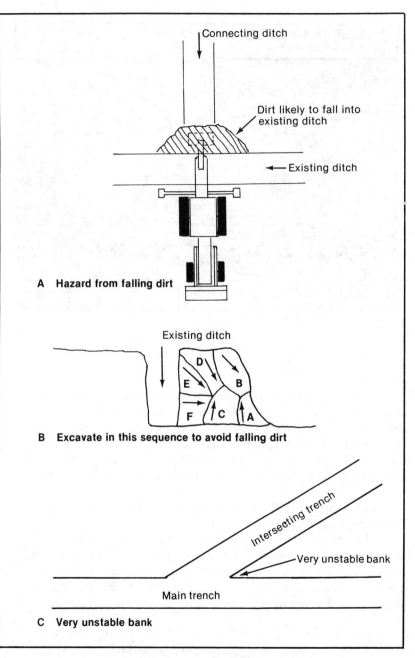

A   **Hazard from falling dirt**

B   **Excavate in this sequence to avoid falling dirt**

C   **Very unstable bank**

Excavating toward an open ditch
Figure 5-16

Courtesy: Allied Steel & Tractor Products

Hole-Hog
Figure 5-17

Steel bars are useful for other tasks also. Some utility line contractors have sets of bars with male and female threaded ends that can be joined together into a single long bar. If soil conditions are right, a bar like this can be punched under the entire width of a street. That can save a lot of time and money, eliminating most excavation and nearly all repairing of the street.

Another useful tool for punching holes is the Hole-Hog shown in Figure 5-17.

And remember, you don't need to dig a 2-foot wide trench for a 3/4-inch or 1-inch water service pipe. A 12-inch or 18-inch bucket will do the job and reduce street restoration. In some cases, the main pipe will be backfilled prior to excavating the water service. This means you'll have to expose the buried main pipe.

Excavating the work area for the saddle or tap— Where a lateral joins the water main, you'll usually have to install a saddle or tap. You cut a small hole in the main line and tap into the main with a special fitting. The danger in this work is that it's done with the main under full operating pressure. If you damage the pipe, you'll not only have to repair it, you'll have a flooded excavation to add to your problems.

To install the tap, you'll need to excavate a work area around the main. If you're using a 1-foot or 18-inch wide bucket, you'll have to widen the trench around the new main. It's faster to dig at 90 degrees to the pipe. But to avoid the chance of damage, consider digging parallel to the pipe. You'll be less likely to damage the pipe, but at the cost of increased street damage. If you decide to dig at 90 degrees to the main, the following technique will help prevent damage.

If the main has just recently been installed, you'll know its exact minimum depth. Allow a 1-foot safety margin all around the new pipe. Don't dig in that zone. Follow along on Figure 5-18. First, dig up to the existing trench line above the known depth of the main. Then excavate up to the old trench line and below the main pipe. Finally, push the dirt down and away from the pipe in the direction shown in Figure 5-18 C. Have the grademan locate the pipe by hand. A little cleanup with the hoe will finish the excavation.

Remember that some pipe is more resistant to damage than others. Small-diameter PVC and transite is very fragile. So spend a little time shoveling if you need to. It's always easier to locate pipe with a shovel than it is to repair the damaged pipe.

Using the Front Bucket

The front bucket on a wheeled backhoe will get plenty of use on most utility line jobs. Wheeled backhoes are both versatile and highly productive machines. But using the front bucket requires an entirely different set of skills. For example, the operator has to understand how bucket loading affects weight distribution on his machine. Pushing the loader into a pile of material transfers weight to the rear of the tractor. Lifting material in the front bucket has the opposite effect. It transfers weight to the front of the machine.

Lifting material in the bucket of a 2-wheel drive backhoe reduces traction on the rear wheels. Carrying a load downhill reduces traction even more. If the grade is steep enough and the load heavy enough, the hoe loses all traction, becoming a runaway.

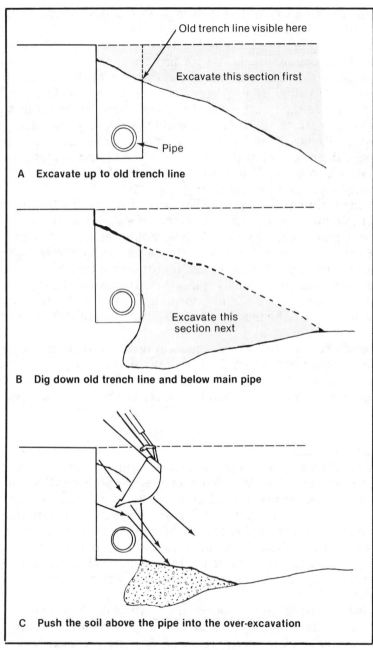

Excavating a service trench to expose buried pipe
Figure 5-18

Avoid using a 2-wheel drive backhoe for heavy material handling and loader work. A 4-wheel drive front loader is better suited for these tasks. Use a backhoe's front loader for smaller tasks when a large front loader isn't available.

A midsize hoe's front loader shouldn't be called on to lift more than about 3,000 pounds. But even with this limited capacity, a backhoe's ability to backfill its own trenches makes it a versatile dual-purpose machine.

You'll handle seven different tasks with a tractor-loader's front bucket: picking up material from a stockpile, picking up spread material, dozing, spread dumping, sprinkle dumping, fine grading and smoothing, and supporting the tractor in muddy ground.

Picking up material from a stockpile— Here's the key to picking up a good bucket load when you don't have the advantage of 4-wheel drive. Charge into the pile at a reasonable speed (about as fast as a brisk walk) and roll the bucket up just as forward momentum stops. That should fill the bucket to capacity.

When cutting into the stockpile, be sure the toe plate and the floor of the bucket are level. Most machines have a bucket angle sight gauge to help the operator judge the angle. Some machines have a plate welded on the back of the loader bucket. The plate is parallel to the bucket floor. This arrangement helps the operator level the bucket or judge the bucket angle.

Picking up spread material— When picking up soil, angle the bucket down so it can cut into the material. As soil builds up in front of the machine, roll the bucket back. When you're picking up the last of the material, approach it at 90 degrees to the previous passes.

Dozing— A backhoe may not be as precise as a grader, but it can still clean up and do minor finish grading. When using the backhoe as a dozer, consider using the independent brakes for direction control — if your backhoe has separate brakes for each wheel. Also note that the foot throttle on some backhoes is located in a position that interferes with operation of the independent brakes.

During normal front loader work, engine power is needed for both hydraulic power and wheel turning power. Using the independent brakes for direction control places extra load on engine power. Watch the engine speed. Don't run your backhoe under load at low engine speeds.

Select a gear low enough so the tractor works without stalling or lugging. Zipping around at high speeds and in high gear makes the operator look efficient. But constant downshifting is hard on the machine and a lugging engine slows hydraulic response. It's better to slow down and work in a lower gear.

Some backhoes have an "instant" forward-reverse shuttle lever. Equipment manufacturers claim this shuttle lever allows you to shift from forward or reverse at full engine speed. Some operators, however, still prefer to use the foot clutch or reduce engine speed when changing direction. This makes for a smoother change of direction.

Use the bucket at a steep angle when you're working on firm ground. This results in a cleaner pass.

Spread dumping— If later spreading is required, it makes sense to spread the material as it's dumped. There are two common ways of doing this. In the first method, move the bucket to about 6 inches above ground level. During travel over the spread area, dump the bucket as you raise the loader. In the second method, make a pass in low gear, simultaneously raising the load and dumping the bucket. This spreads the soil while it's being dumped. See Figure 5-19.

Sprinkle dumping— Use this method for accurate placement of granular material such as sand or gravel. Select a low gear. Raise a full bucket load to a height above the operator's eye level so the operator can see material as it falls from the bucket. Dumping starts as the operator approaches the spread area. Use the dump control lever to shake the material out of the bucket during travel. An expert operator can spread a fairly even layer this way. Note, however, that quick hydraulic response is required. Keep the engine speed up.

Fine grading and smoothing— There are several ways of smoothing dumped material with a backhoe. The first is to use the bucket in a slightly rolled-back position to doze the

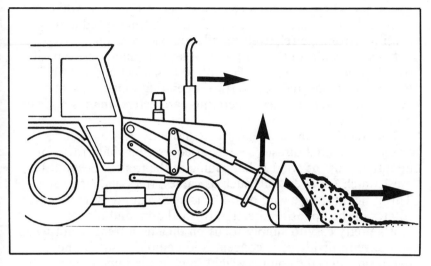

Spreading the load while dumping
Figure 5-19

material. The second is to drag the bucket backward so the cutting edge of the toe plate or the angled rear edge of the bucket floor smooths the material.

A backhoe doesn't do very well as a fine grader because there's no way to adjust the horizontal angle of the front bucket. If the rear wheels aren't level, the front bucket won't be either. If your front bucket isn't level, then your fine grading won't be very precise. The only solution is to level part of the area with the hoe, then use this level area to work from.

Supporting the tractor in muddy conditions— Use the front loader to support the front of the tractor when working in mud. This is the same support system as we used for moving across an open ditch. Look again at Figure 5-15.

The Tracked Backhoe

The tracked backhoe (or trackhoe or excavator as it is sometimes called) is a much more difficult machine to operate than the wheeled backhoe. In my experience, it's the older, mature and experienced men who make the best trackhoe operators. These operators can sometimes "hot-dog" while operating the smaller wheeled backhoes, getting speed and precision from the relatively small, compact machine.

But they can't do this with the larger machines. The typical trackhoe has a horizontal reach of 30 to 40 feet and the ability to revolve 360 degrees. Its vertical operating range is perhaps 50 vertical feet. Within these operating parameters, there's a great potential for destruction.

The trackhoe operator must be constantly aware of three areas of potential disaster. First, of course, his main focus of attention must be what is happening in the trench. Second, he must watch for overheads such as cables, trees and structures. Third, he must be aware of the machine's counterweight, which rotates in the opposite direction of the boom and dipper.

It's pretty easy to simply focus on digging a trench. But when you have to divide your concentration among a safe trench, overheads, and the counterweight, and at the same time watch for damage to crosslines, *and* maintain a profitable production rate, you've got a real job on your hands. This isn't for the inexperienced operator.

In a typical urban sewer installation project, the trackhoe operator will be dealing with all of these pressures 8 to 10 hours a day. To further complicate matters, the hoe may be used to install heavy pipes, place the pipe bedding, move shoring, move dewatering equipment, load and work around trucks, and coordinate with auxiliary equipment such as loaders or dozers.

Attention to Safety Is a Must

Safety must be stressed by everyone on the job site. On one job site, the operator, concentrating on what he was doing with the hoe, forgot the counterweight. He snapped off a telephone pole. If the pole had carried electricity, someone could have died. In another instance, a small cave-in trapped a worker in a trench; the trackhoe operator then broke a large diameter water main, and the trapped worker drowned in the flooded trench. If you've worked around heavy equipment, you probably have horror stories of your own.

The potential for injury and death is ever present. Injury from rock or spoil falling from an overloaded hoe bucket or a spoil pile too close to the trench is a real, but avoidable, hazard.

The responsibility for maintaining a safe work place should be shared by the whole crew. The ground crew can see whether or not the counterweight will clear an obstruction. They can make it a point to remain in sight of the operator, alert for problems and signaling requirements.

The competent pipe crew and operator will have a system of hand signals to communicate over distance and noise. These hand signals aren't standardized, but in most cases a closed fist means *stop*, and an open hand with the thumb and index finger forming a circle means *OK*. Circling a vertical index finger means *raise*. Showing the operator a certain number of fingers pointing up or down means to move that many tenths of a foot up or down, while pointing to your foot means the measure is a foot.

Every crew will have variations and additions to these hand signals. It's not important which signals your crews use, as long as *every member of the crew* understands them and uses them to communicate visually. Shutting down to give directions or clear up misunderstandings isn't good for safety *or* production.

Selecting the Trackhoe Operator

Your trackhoe operator is one of your most important crew members. If he lacks experience in deep utility installation, his experience in other kinds of equipment or in trackhoes in other applications is largely irrelevant. And even if the operator has the experience, avoid anyone who's immature, a showoff, or just plain unstable. He'll cost you money and maybe hurt or kill someone.

It's part of the operator's job to anticipate changing conditions and allow for them. It's not simply operating a piece of equipment. The job entails awareness of hazards, planning spoil placement, and planning the job so the subsequent work can be efficiently and safely accomplished.

For instance, when wet sandy spoil from the bottom of a trench is piled on the banks, the water will drain out of the wet spoil quickly and seep down into the banks. This added water will lubricate the soil in the bank, possibly causing the otherwise stable bank to slip. An experienced operator should be able to foresee this, and avoid it.

It's a job where experience, maturity and safety consciousness are as important or more important than the mechanics of operating the machine. This point is often lost on young, ambitious operators. There's a saying in utility work: "If you ain't smart enough to operate a number 1 hand shovel, you sure

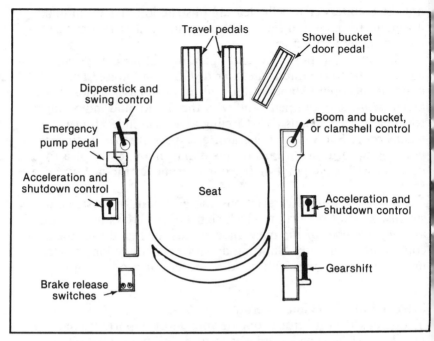

Cab layout of tracked backhoe
Figure 5-20

ain't smart enough to run a hoe!'' That means that the best possible operator experience is ground work, pipe laying or other work around the crew. Having worked in a 20-foot deep trench will increase an operator's safety awareness a lot faster than any safety program, book or lecture.

Safety on the job is a subject I feel strongly about, but my lecture's over. Let's see when you should use a trackhoe, and how to use it to make money on an efficient job.

When to Choose the Tracked Backhoe
For trenches deeper than 8 feet, or in extremely hard digging, the tracked backhoe will be your first choice. It's designed especially for deep trenching and for truck loading. With the right bucket, a tracked backhoe can also be used for material rehandling and site cleanup. Figure 5-20 shows the cab layout of a tracked backhoe.

Some contractors avoid using tracked backhoes because they're big and expensive to operate. And if your operator is slow or inefficient, it's even more expensive. But remember that it's the capacity of a tracked backhoe that makes it so productive. And while the hourly or daily cost may be high, a skilled operator can *save* you money by moving large quantities of dirt in a short period of time. Used wisely, a tracked backhoe can increase your production rate and your profits.

Tracked backhoes differ from wheeled backhoes in their size, equipment rotation, field of vision, counterbalance, extra-sensitive controls, and cycle times.

• Size: Some tracked backhoes are so big that the operator can look in a second-floor window while seated in the cab. That makes it more likely that a careless operator will damage people and property accidentally.

The boom and dipper can reach high voltage cables with ease. And note this carefully. You don't have to actually touch an electric line to attract a deadly charge. High voltage electricity will jump several feet from the cables to the backhoe. When this happens, the machine conducts electricity to the ground, possibly electrocuting anyone nearby if the ground is flooded or saturated. Keep your equipment at least 10 feet from high voltage power lines.

• Equipment rotation: Wheeled backhoes can only rotate 180 degrees. Tracked backhoes can revolve a full 360 degrees. The operator's cab revolves with the turntable.

• Field of vision: Vision is restricted on the boom side of the cab. The operator's view to the rear (the direction of travel) is also severely restricted. Forward vision is usually excellent.

On machines with the cab set on the left, the operator's blind spot is usually on the right. Cabs set on the right side of the machine have a blind spot on the left.

Working close to power poles, trees or existing structures can be hazardous. Have a spotter on the ground help keep the hoe clear of obstacles. Once the operator makes a mental note on the position of an obstruction, he won't need the spotter again until he repositions the hoe.

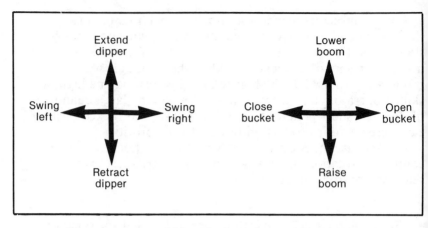

Common control lever layout on tracked backhoes
Figure 5-21

• Counterbalance: Tracked backhoes, unlike wheeled machines, have counterweights. When you swing the hoe left, the counterweight at the rear of the machine swings right. It allows the machine to handle heavy loads. But it's an additional area of potential hazard. The counterweight swings with enough force to down power poles or damage vehicles and buildings.

• Extra-sensitive controls: The controls on a tracked backhoe are extremely sensitive. You'll hear operators say the machine is "jumpy" because it's so responsive. Twin joystick controls are most common, but some machines have combination hand and foot controls.

European countries have standardized the hoe control layouts, but in the U.S. there are still several different control layouts used on tracked excavators. Back in Figure 5-1, we illustrated the twin-lever layout found on many wheeled backhoes. Some tracked backhoes use it, too. Figure 5-21 shows the layout found more often on tracked backhoes, with the dipper and bucket controls in different hands. This pattern was popularized by its use on Caterpillar model 215, 225, 235, and 245 tracked excavators.

Different types of lever configurations work well under different conditions — and each design has some advantages. But the best control configuration for any job will be the layout our operator knows best. It takes time to learn precise maneuvers when doing complicated work. The operator has to learn what combination of moves will give him the desired response. And he has to learn these combinations so well that reactions become automatic, as though the machine were an extension of his body.

An experienced operator doesn't have to watch the panel. He doesn't even think about which movements are desired. He just senses the correct combination of motions. You can guess what will happen if you switch machines on him. His perfectly tuned hands will try to go through all the usual moves — moves which won't work on the different lever configuration.

Like most operators, I'm inept at operating a hoe fitted with an unfamiliar control layout. Imagine driving a car with the clutch and brake pedals switched around. You'd certainly be hitting the wrong pedal until your habits slowly changed. It's the same with a hoe operator. Changing the controls means having to discard old habits and form new ones.

Any time you put an operator on a unfamiliar piece of equipment, give him a little training time. Let him practice on simple work in an open area, or loading trucks, before he takes on more complex work. After a little practice the movements will become automatic again. Also, you should know that most control lever layouts can be altered to suit the operator.

Cycle times: The larger the machine, the slower the cycle time. For example, a wheeled backhoe has a minimum effective cycle time of about 10 seconds. A small 3/4-yard tracked hoe has a minimum effective cycle time of about 15 seconds. Large tracked excavators take as long as 30 to 40 seconds per digging cycle.

A slower digging cycle makes it very important that the operator get a full bucket load with each cycle. A good operator fills the hoe bucket to capacity at least 90 percent of the time.

Now that we've identified the distinguishing features of the tracked backhoe, let's look at the basic digging technique for this machine. Then we'll discuss five specialized tracked hoe tasks you'll be performing in your utility line work.

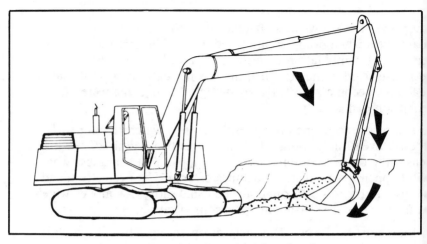

Curling force of tracked hoe bucket
Figure 5-22

Basic Digging Technique

The key to digging with a tracked excavator is *feathering* the controls. The operator's small, precise movements are amplified many times by the hydraulic system. Fluid under pressure flows to the rams. The speed and movement of each ram is determined by the speed and direction of fluid flowing into each cylinder. Clumsy control movements force a lot of fluid into the ram suddenly. The result is sudden, jerky movement.

On a wheeled backhoe, the operator often pulls the control levers wide open, allowing the hoe to operate at maximum speed and power. A tracked hoe operator will sometimes do that, but only when raising the boom of his machine. He'll usually use several control functions simultaneously to slow the equipment, aiming for smooth precision. Remember that if you overextend a tracked excavator's ram by an inch, you'll overextend the bucket by about a foot.

Digging technique for a tracked backhoe is slightly different from wheeled backhoes. On a wheeled backhoe, the operator usually rakes the ditch bottom, scraping off soil to fill the bucket as it's pulled toward the operator. The great power of a tracked hoe lets the operator scoop out full bites at a time.

As you move the dipper around the radius of the bucket, curl the bucket to scoop up dirt. If you lower the boom as you do this, it will add to the curling force of the bucket. See Figure 5-22. This technique works especially well in clay.

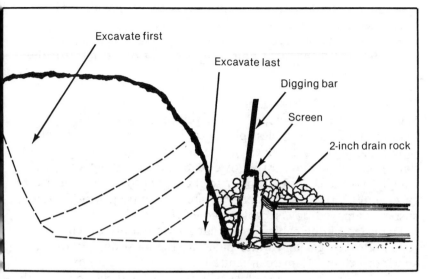

Excavating in groundwater
Figure 5-23

When spreading bedding, the tracked backhoe normally uses a raking technique. Skilled operators can actually *throw* bedding rock from the bucket. Here's how it's done. As you extend the dipper, drop the boom and open the bucket. This is faster than sprinkle dumping. The control layout shown in Figure 5-21 works well when throwing bedding. This layout has the bucket and dipper controls in opposite hands, making the maneuver easier.

Excavating in Groundwater

Figure 5-23 shows how to excavate in high groundwater. You want to prevent water and mud from flowing against the pipe that's already been laid. To do this, start excavating some distance behind the end of the pipe, leaving "dam" of unexcavated soil until the last. This helps stop mud and sludge from blocking the pipe.

Material Rehandling and Cleanup

A tracked backhoe fitted with the right bucket can be used for cleanup and material rehandling. It's possible to use the swing rams to sweep material sideways. But this has to be the least efficient way to do cleanup. And it stresses the machine. Do it only when no other equipment is available.

You could use a larger bucket, but it probably wouldn't be cost effective to change it unless there's a lot of cleanup for the machine to do.

Working Around Cross-Lines

When working in developed areas, you're going to find underground power cables, phone cables, water pipes and gas pipes. Damaging these lines can be both expensive and dangerous. When you hit a power cable, circuit protection fuses usually blow and cut off power to the damaged cable. But if the fuses don't blow, you've got live wires in the trench. So if you suspect damaged power cables, move with caution.

Damaged phone cables aren't dangerous but are expensive to repair. Broken gas and water pipes are the most dangerous. A spark or exhaust flame can ignite the gas, creating an explosion and fire. Ruptured water pipes will quickly flood the trench and can easily drown a trapped worker.

Cross-lines present a special problem when you're working with a tracked backhoe. Large tracked hoes often dig very wide trenches. Any cross-lines will be suspended across this wide trench, especially if the line crosses the trench diagonally. Some are fragile enough that they may collapse under their own weight. Occasionally a line will snap when the trench is backfilled. Transite, clay, small-diameter PVC pipe and concrete pipe and duct are especially fragile.

Support fragile cross-lines if the trench is wide. Lay a stout 8-inch by 10-inch timber across the ditch. An old utility pole works well. Suspend the cross-line from the timber with chains or slings until the work is complete. Figure 5-24 shows a transite water line supported in a sling. After the pipe layers have passed under the cross-line, compact fill under it by hand to reduce sag and the chance of breakage. Use a fine, easily compacted material like sand to fill under fragile cross-lines.

If your trench is deep and your cross-lines are fairly shallow, a tracked backhoe can be used to good advantage. The long reach of the tracked backhoe lets it undermine shallow cross-lines with ease. Once you've tunneled under the line, push the unsupported dirt down and away from the cross-line.

Have a grademan direct the operator any time you're working around a cross-line. Using hand signals, he can direct the operator's movements around the obstacle.

Courtesy: Dick Zanker

Supporting a cross-line in a sling
Figure 5-24

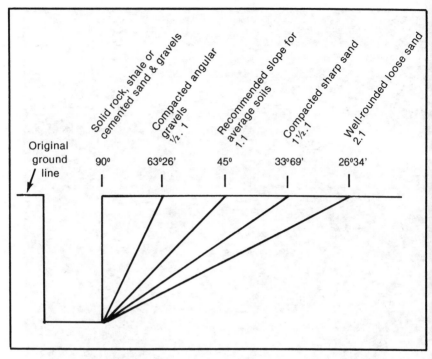

Angle of repose for common soil types
Figure 5-25

Working with Shoring Systems
It's dangerous and illegal to have anyone working in a trench that's over 5 feet deep unless the trench is protected with some kind of cave-in protection. Of course, the danger of collapse varies with the soil type. Some cemented soils are quite stable. But it's wise to consider all trench walls over 5 feet high to be dangerous. Any trench can collapse and bury your crew. Don't let that happen on your jobs.

The easiest way to protect against cave-ins is to slope the trench walls. Figure 5-25 shows the *angle of repose* of several common soil types. This is the angle at which the soil will rest. But don't rely on a chart to tell you when the soil is safe. There are other factors to consider:

• Damp or saturated soil is much less stable than dry earth.

• Changing soil conditions can make a trench wall unsafe. A vertical trench wall may be quite safe when dug in cemented

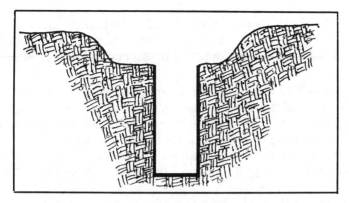

Benching
Figure 5-26

gravel. But if you run into a pocket of loose sand or heavy oam, the wall becomes dangerous.

• Earth that's disturbed can become unstable. For example, equipment digging another trench or vibratory compaction equipment nearby may make your digging area unstable.

Safety is always your top priority. No job, no completion schedule, no profit margin is worth the life of a crew member. But making every trench completely accident-proof isn't practical. In difficult conditions, sloping an open trench enough o make it safe can take a lot of time. When open trenching is mpractical or impossible, select one of the three common shoring systems: shoring jacks, steel sheets or the trench box.

Shoring jacks— Shoring jacks are the best choice when working n firm soil that will stay in place long enough for the jacks to be positioned. They're easy to use. You can place and remove hem by hand without entering the trench. As the excavation proceeds, the pipe crew moves the jacks forward.

If you use shoring jacks, the equipment operator has to keep he trench walls vertical and smooth. Figure 5-26 shows a trench hat's protected at the top by benching and sloped banks. Trench walls are vertical and smooth at the bottom so shoring acks can be inserted. This excavation technique is known as *benching*. It lets the pipe crew position the jacks easily and accurately.

Steel sheets— Steel sheets are bulky and have limited use. But they're fairly simple to install and remove. Shoring jacks placed between 1-inch thick sheets support the trench walls.

Trench box— The most difficult shoring system to work with is the trench box, also known as the *coffin.* It's the best shoring when working in very unstable ground. The trench box uses less labor than shoring jacks but requires a skillful operator to maneuver the box quickly and safely. It also takes a large, powerful trackhoe to pull the heavy box that's squeezed by the trench walls.

Pipe layers work inside the box while the hoe bucket is working in and immediately in front of the box. That can be dangerous. The hoe operator must work carefully, moving the bucket only when he's absolutely certain that no one is in the way.

There are two ways to use a trench box. The technique you select will depend on whether you're working in stable or unstable soil.

• Stable soil: When working in stable soil, use the method shown in Figure 5-27. Pull the box ahead as the trench is extended. Here are the five steps to follow:

1) Excavate a section of trench, beginning with an open cut down to pipe grade level.

2) Lower the trench box into the trench.

3) Install the pipe. See Figure 5-27 A.

4) Excavate ahead of the box for the next length of pipe as shown in Figure 5-27 B.

5) Pull the box forward into the new excavation while backfilling behind it. See Figure 5-27 C.

• Unstable soil: When working in unstable soil, follow Figure 5-28. Pull the box up so it rests over the area to be excavated. Trench between the box walls. Force the box down into the newly dug trench as excavation proceeds. This is slow work. But it may be the only thing that works in unstable ground.

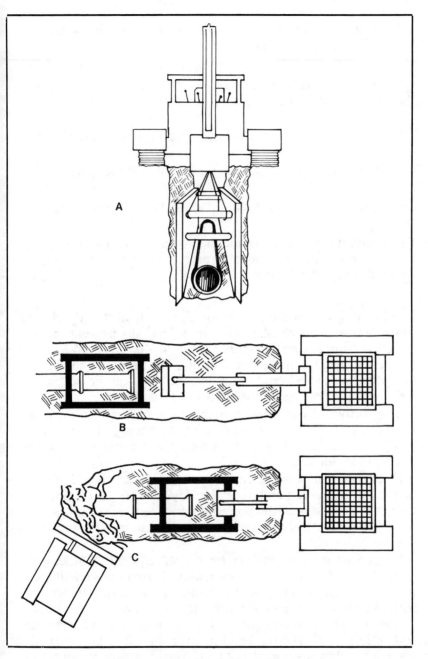

Using a trench box in stable soil
Figure 5-27

Here are the five steps when using a trench box in unstable soil. Follow along on Figure 5-28.

A)   Move the box over the spot where the trench will be. Dig the trench from inside the box.

B)   Tamp the box down after each bucket of soil is removed.

C)   When the box is at the right grade, install the pipe.

D)   Pull the box forward and up at about a 45-degree angle.

E)   Continue excavating inside the box and tamping it to grade so you can set the next length of pipe. As you pull the box forward, backfill behind it.

When positioning the trench box, it's essential that the box be centered over the pipe. And be sure the box sits high on the trench walls as shown in Figure 5-29. If the box sits too low in the trench, pulling the box may disturb the bedding and the newly laid pipe. PVC and small diameter pipe can be pulled apart when a low trench box is moved forward.

The bottom of the trench box should sit higher than the pipe — about 2 to 3 feet higher will usually be enough. But don't position it too high. If it's higher than 5 feet above the pipe, the ditch will be unsafe.

In extremely unstable ground, your pipe layers should climb in and out of the trench box as they lay each length of pipe. Don't let them stay in the box while the hoe is excavating soil or moving the box. It's common to have dirt fall into the box if the soil is unstable.

Using a trench box cuts the excavation required by at least a third. Figure 5-30 shows the soil removed from the same trench when excavated as an open trench (no trench box), a trench with one trench box and a trench with two trench boxes. The savings shown are based on a trench assumed to be 12 feet long and 18 feet deep. The soil condition is assumed to be stable enough to allow a slope of 1/2 to 1. The excavator is using a 2-cubic-yard bucket. Chapter 8 has more information on shoring.

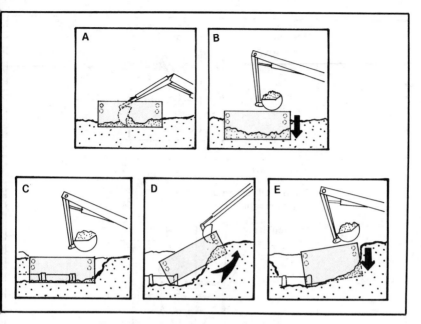

Using a trench box in unstable soil
Figure 5-28

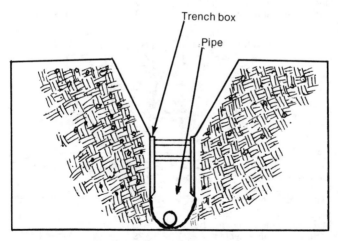

Positioning the trench box
Figure 5-29

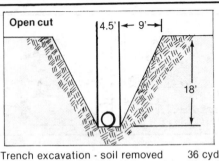

Trench excavation - soil removed	36 cyds
Layback - soil removed	72 cyds
Total - soil removed	**108 cyds**
Total - machine cycles	**54**

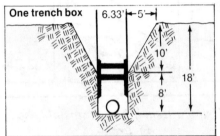

Trench excavation - soil removed	50 cyds	
Layback - soil removed	23 cyds	
Total - soil removed	**73 cyds**	**saves 32%**
Machine cycles - excavating	37	
Machine cycles - moving box	3	
Total - machine cycles	**40**	**saves 26%**

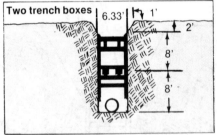

Trench excavation - soil removed	50 cyds	
Layback - soil removed	1 cyd	
Total - soil removed	**51 cyds**	**saves 52%**
Machine cycles - excavating	26	
Machine cycles - moving box	3	
Total - machine cycles	**29**	**saves 46%**

Spoil savings
Figure 5-30

Working in Mud or Soft Ground

Tracked backhoes can work in several feet of slimy mud when necessary. Still, I've seen tracked backhoes get mired down in deep mud. Here's what to do if a tracked hoe gets stuck.

Use the hoe to lift and pull or push the machine out of a wet spot. If depressing the travel pedals on your hoe leaves the hoe without any hydraulic power, feather the travel pedal. This reduces the power needed for travel so more is available for the hoe. In extremely boggy conditions, you may need heavy timbers or excavator mats. If this doesn't work, a dozer may be needed to pull the machine out.

6

SOIL COMPACTION

Underground utility contractors obviously do a lot of excavation. And every trench they dig has to be filled back up after the utilities are installed. This seems like a simple enough task. There's nothing easier than digging a hole and filling it back up again, right? If you believe that, you haven't been in utility line contracting very long. Backfilling a trench can be a very complex and expensive operation. This chapter is intended to help you avoid some of the more common (and more expensive) mistakes.

Most soil in its natural state is pretty stable — at least when it's dry. It's probably been resting undisturbed for hundreds, thousands or even millions of years. But when your blade or bucket cuts into it, all that's changed. It loosens up and *swells* because you introduce air cavities into the mass. Loose soil is unstable and has a low load-bearing capacity. As backfill, it shrinks and settles, probably causing at least some damage to surfaces. To protect against settling, the specifications for your job probably demand that you compact all backfill, restoring it as nearly as possible to its previously undisturbed density, or even greater.

In this chapter, I'll explain what's required to protect your backfill against settling. First we'll look at the structure of soil, why it swells and why it compacts. Then we'll look at soil density and how the density of soil is tested. Finally, we'll look at the most efficient methods of soil compaction.

Soil Structure
Soil consists of rock particles, air voids and water. The soil particles vary in size from the very fine, in silt or clay, to the larger particles of sand and gravel. Under a microscope, you can

see the particles resting against each other like rocks in a rock pile. They have irregular shapes and don't fit together very well. Spaces between soil particles are air voids. These voids decrease soil density and make it unstable. More air voids are introduced any time you disturb or excavate soil. Air voids mean looser soil and less stability. Reducing air voids is the goal of soil compaction.

Here's what happens when you compact soil. Imagine you've put some disturbed soil in a press. The press crams the soil particles closer together. You'll reach a point where the particles are as close as they can get. They can't get any closer because of their irregular shapes. But there are still air voids. Continued pressure would break down the soil particle and result in a fine powder that still contains air.

But if you add water to this soil sample, it does two things. First, it "lubricates" the particles so they slide together easier when you compress them. Second, water fills the remaining air voids.

Water makes compaction easier. But how much water is needed? Too little water won't lubricate the particles enough or fill all the voids. Too much water will hold the particles apart, causing them to float. If you try to compact water-saturated soil, the soil will *pump* (become unstable) and move away from the compaction equipment.

The right amount of moisture for compaction is called the *optimum moisture content*. Every soil has its own OPM. Determining the soil's optimum moisture content is critical. It's usually expressed as a percentage of maximum dry weight.

This point is important enough to restate: *It's virtually impossible to compact soil without adequate moisture — or to compact soil containing too much water.*

Determining Soil Density

The closer the soil particles, the more dense soil becomes. Reducing or eliminating air voids increases the soil's density and its load-bearing capacity, minimizing the chance of future settlement. The whole point of soil compaction is to increase soil density, essentially eliminating the air voids.

Density can be determined as a weight-to-volume ratio of the soil. The more soil particles per cubic foot, the greater the weight per cubic foot and the greater the density. Three methods for testing soil density are the Proctor test, the sand cone test, and the nuclear density meter.

Proctor test— In the 1930s, R. R. Proctor discovered the important relationship between the soil's density, its moisture content, and the amount of compactive effort needed. He discovered that when he varied the water content in a soil, equal compactive effort would result in different weight-to-volume ratios.

It's important for you to understand the concept of the Proctor findings. Applying a heavy weight to a perfectly dry soil sample won't achieve much increase in density. When water is added, it's much easier to increase the density. Remember, the water acts as a "lubricant."

To understand this concept, visualize a pile of dry flour and a lump of dough. What happens when you apply pressure to each? The dry flour displaces easily. But when water is added to form a stiff dough, it's much more malleable and can be compressed. If too much water is added, it becomes sloppy and again displaces easily. Here's the point: It's impossible to compact a soil unless it contains the correct proportion of water.

The Standard Proctor test process reveals this clearly. First, a sample of the soil is dried in an oven to remove all the water. It's weighed, then a small amount of water is added back to it. Part of the sample is then placed in a mold with a volume of 1/30 cubic foot and subjected to a series of blows. Specifically, a 5½-pound hammer is dropped 25 times from a height of 1 foot, for a total compactive effort of 12,400 foot-pounds. Figure 6-1 A shows the process. Then second and third layers are added to the mold and compressed. Finally, the mold and compressed sample are weighed. When the weight of the mold is subtracted, the result is the weight of the soil sample.

Now the tester knows the volume (1/30 cubic foot), the weight, and the amount of water added to this particular sample. The next step is to compute the cubic foot weight. Then the whole process is repeated, each time adding extra water to the sample. Like dry flour, the soil changes with added water. Let's assume that initially, when it was too dry, it weighed 100 pounds per cubic foot. Additional water increases the weight/volume ratio, until too much water is added and the weight/volume ratio begins to decrease. This was R.R. Proctor's important discovery. *To attain maximum density, soil must have the correct moisture content.* There's also a Modified Proctor test, shown in Figure 6-1 B. The process is the same, except that

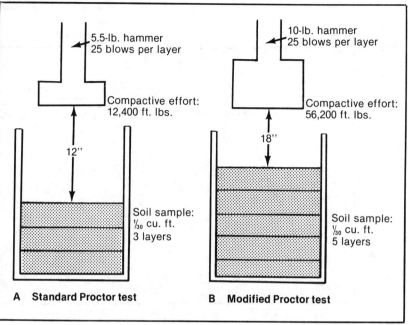

Proctor tests
Figure 6-1

the samples are compacted in five layers with a 10-pound
hammer dropped from 18 inches. This is a much greater
compactive effort; 56,200 foot-pounds instead of 12,400.

Most airfield runways, military bases and Federal highway
projects specify the use of the Modified Proctor as the standard
of comparison for compaction results. Most contractors agree
that the Modified Proctor is a very tough specification to meet.
If job specifications demand 95% compaction using the
Modified Proctor as the comparison, you'll face compaction
problems unless you're experienced and have access to very
heavy compaction equipment.

Figure 6-2 shows a typical density curve plotted as a result of
a Proctor test. Point 1 on the curve shows the weight of the
sample with an 8% moisture content. As the moisture content
increased, the weight increased — until too much water was
added and the curve started back down. For this particular
sample, the point of *100% density* was achieved at a moisture
content of about 11%.

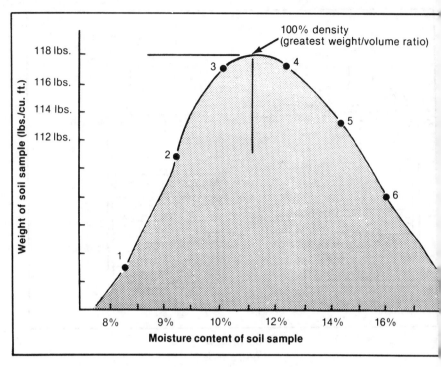

Curve showing 100% density and optimum moisture content
Figure 6-2

Compactive effort at the optimum moisture content is essential for efficient compaction. If the soil is much above or below the optimum moisture content, you'll never get the right soil density. If the soil is too wet, it must be allowed to dry; if it's too dry, water must be added.

Remember that each soil type has a different weight-to-volume ratio and different optimum moisture content. The Proctor test finds the point taken as 100% density and optimum moisture content for the soil you're compacting. Each soil type will produce a different curve because each one has a different optimum moisture content.

Sand cone test— This test compares the density of a field sample to the density of the Proctor test sample. The field tester takes a precise volume of soil from the job site. He weighs the sample dry and records the weight per cubic foot. Using the weight-to-volume ratio at 100% density from the Proctor curve, he can now calculate the density percentage of the field sample. Let's look at an example.

Assume the Proctor test shows 100% density when a cubic foot of soil from your site weighs 125 pounds. By deducting the moisture content, the technician calculates that the dry weight would be 111.6 pounds. Now the technician takes a precise volume of soil from your backfill and compares it to the Proctor sample. Let's say the field sample weighs 107 pounds per cubic foot dry. Convert these numbers to a percentage of optimum compaction: divide the dry weight of the field sample by the dry weight of the Proctor sample and multiply by 100. Your calculation will look like this:

$$\text{Density percentage of field sample} = \frac{\text{dry weight of field sample}}{\text{dry weight of Proctor sample}} \times 100$$

$$\frac{107 \text{ lbs./cu. ft. dry}}{111.6 \text{ lbs./cu. ft. dry}} = .96 \times 100 = 96\% \text{ density}$$

The disadvantage of the sand cone test is that it takes time. There's a newer method that's faster: the nuclear density meter.

Nuclear density meter— When gamma rays are directed at any mass, some pass through the material and some are reflected back at the gamma source. The density of the material affects the number of rays reflected. The nuclear density meter releases gamma rays into the soil and records very precisely the rays that are reflected. Soil density is calculated from the meter reading.

For example, assume the machine emits 25,000 rays. Let's say 22,000 rays are returned to it. On this particular machine, 22,000 rays indicates a dry weight of 122 pounds per cubic foot. A Proctor test on the same soil may reveal 100% density at 125 pounds per cubic foot dry weight. That indicates over 97% compaction (122 divided by 125), which would be acceptable under most job specifications.

When the nuclear density meter is used, here are three important facts to remember:

• Gamma rays reflected show the *average* density of material under the probe. If there are layers of high density and layers of low density under the probe, it may be hard to determine which you're measuring. For example, suppose the probe is set on soil that rests 8 inches above a very dense object like a rock or boulder. An artificially high count of gamma rays may be returned. If the probe is set on soil that's 8 inches above a very loose sublayer, some rays may dissipate into the loose soil and not return to be counted.

• If soil tested happens to be above the optimum moisture content at the time of testing, the excess moisture is deducted from the score. That makes it harder to get an acceptable compaction score. Be sure the soil isn't above optimum moisture when you take the test.

• A dried crust on the surface of the test area may reduce the count of returned rays. This will give a higher weight to volume reading than if the site were freshly excavated.

As with most other tools and instruments, the accuracy of the nuclear density meter depends on the skill and experience of the user. Some users, for example, have found that a nuclear density reading taken in the bottom of a narrow trench may not be accurate. And as with any meter, the density meter must be regularly calibrated by testing it on a material of known density.

Compaction Methods

Soil type, required density, and moisture content all affect your choice of compaction method. Sand and gravel compact well with water and vibration. Silt and clay may require a precise combination of vibration, weight and moisture. The two most common methods of compaction are water settling or *puddling* and a combination of weight and vibration.

Puddling
This is the easiest compaction method. Simply flood the loose backfill with water until it's completely saturated. As the water

floods the soil, air voids are driven to the surface. As the water drains or evaporates, it tends to draw the soil particles together.

An ocean beach offers a good example of how water compacts sand. Above the high tide mark, sand is dry and loose under foot. Just below the high tide mark, the recently wet (but already drained) sand is damp and firm under foot. Below this area, sand is still being washed by waves and is saturated to the point where it's unstable.

The success of the water-settling method depends on the soil type. This method is least effective in silt, clay and other soils that take a long time to drain. Water settling is most effective in sand, gravel and desert-type soils. When used correctly, you can sometimes get compaction up to 80% of the Proctor density. To get more than 80% density, you'll have to use mechanical compaction in addition to puddling.

If you have fairly shallow trenches and hot weather, flood the trench, wait until the soil dries to near the Proctor optimum moisture content, then mechanically compact. If you compact when the soil is too wet, the soil will pump away from your compacting equipment. This is not so much the case with coarse sand and gravels, but you'll still get the best results when the soil is near optimum moisture.

Whether you use water settling alone or puddling and mechanical compaction, let the soil dry out before testing the density. If you're working in sandy soil, you won't have to wait long. With silt and clay, you'll have to be patient. They take longer to dry out. But don't jump the gun — if you test too soon, the density won't be high enough to meet construction standards.

Because the subject of soils and soil engineering is so complex, it's hard to establish firm guidelines. Course-grained soils tend to withstand saturation very well. They drain quickly, and can be worked while they're still quite wet. Fine-grained soils behave very differently. They'll only compact when they contain a very precise moisture content. A fine-grained soil that's poorly compacted will be stable *until* the day it becomes saturated. Then it may collapse and sink dramatically. Even if you don't have to satisfy compaction specifications, it's important to pay attention to good compaction practices.

This means close supervision and adequate training of your crew. Backfill material isn't simply piles of dirt. It's a specific type of soil that requires precise treatment and processing. By

Courtesy: RayGo

Sheepsfoot vibratory roller
Figure 6-3

training and necessity, your equipment operators are dirt movers. Their job is to move great quantities of earth quickly and efficiently. But you've got to make them understand that in building compacted fills, the treatment and processing of the fill material is as important as the quantity moved. They've also got to understand that a process that worked well on one job or on one area of a large project won't necessarily work on the next.

Successful compaction depends on your skill, experience and judgment. Use the data generated by passing or failing the tests to find the best compaction methods. And pay attention to changing conditions. There probably won't be one final answer for the whole job.

Weight and Vibration
Pressing down on most soils will increase the density, especially when weight is applied with high-intensity vibration. Vibration tends to shuffle the soil particles together as the weight compresses them. The two most commonly used tools for this method are the sheepsfoot vibratory roller and hand-held or machine-mounted vibratory plates. See Figures 6-3 and 6-4.

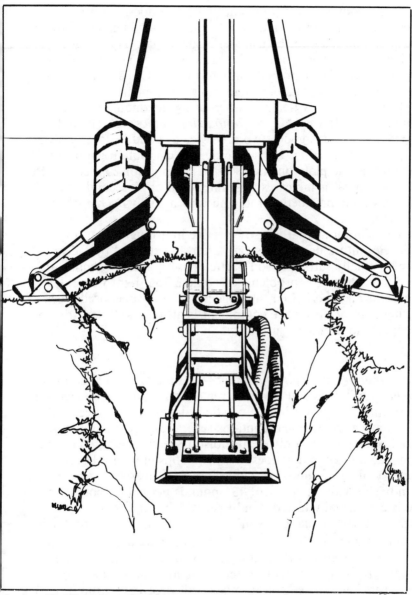

Hoe-mounted vibratory plate
Figure 6-4

If the compaction equipment you select doesn't get the desired result within a reasonable amount of time, something is wrong. You may not be using enough weight. Try a bigger roller or a trackhoe-mounted vibratory plate. You may be trying to compact a layer of soil that's too thick. Or the material may be above or below its optimum moisture content.

In rare cases, particularly when using hoe-mounted vibratory plates, the odd case of *overcompaction* can occur. The old saying: "If one is good, two are twice as good," doesn't always apply in compacting. Let's assume you can obtain good results by placing the plate on each section of spoil for 20 seconds. But if your operator increases this time to 30 or 40 seconds, he decreases the production rate and may actually *decrease* the soil density.

The vibrations that are so effective in arranging the soil particles into a dense mass can, after an extended period of time, actually rattle the soil particles apart again. Complexities like this add to the problems of compaction, reinforcing the need for an experienced, intelligent and well-supervised crew. Using high-performance equipment is all very well, but it's your people who'll make you money.

Soils that are difficult to compact— There are two situations where you'll find soil that's extremely difficult to compact. One is soil that's too wet in its undisturbed state. The other is when you have a job site that contains several different soil types. Each soil type has a different moisture requirement, so a single compaction technique doesn't work. Topsoil with a Proctor density of 110 pounds per cubic foot may overlay a heavier sandy soil with a density of 135 pounds per cubic foot. If these soils are partially mixed, you'll get widely varying density results. In this situation, you'll need to get a Proctor for each soil type and mix of soil types. Try to avoid having to do this by keeping the soils separate. Then you can compare apples to apples (topsoil to topsoil Proctors, sandy soil to sandy soil Proctors and mixed soil to mixed soil Proctors) without repeatedly running the lab test.

Compacting imported fill— Where the excavated soil is known to be unsuitable for fill, imported fill will be specified in the contract documents. In many ways, this is easier. You know

n advance the exact cost of labor, equipment and materials
needed to import fill, and can predict the compaction
characteristics in advance. Compacting native fill always
involves some unknowns.

Excavating with limited working room— Lack of space
complicates the compaction operation. If you're on a job where
there's a lot of earthwork going on, you'll probably have plenty
of room to mix soil, and to add and subtract moisture as
required. If you're working in an urban area, you won't have
this advantage. Dirt excavated from one section of trench may
have to be returned to that section — and often on the same
day it was excavated. Limited working room means you won't
have room to mix soil. It may also mean you can't easily use the
water necessary for wetting. In a narrow street, it's virtually
impossible to wet the soil with a 4,000-gallon water tanker.

The best way to wet backfill in a narrow right-of-way is to
use 2-inch flexible fire hose tapped onto a fire hydrant or large
water tanker. As the material is excavated and backfilled, have
it sprayed with water. The hose man and the equipment
operators must understand the process of evenly wetting the
material, and carefully coordinate their work to obtain the
desired results.

As the spoil is placed back in the trench, it must be spread
into layers called *lifts.* The lift may need to be as thin as 6
inches in difficult-to-compact material that has to meet a tough
specification. In easily compactible material, on the other hand,
several feet of backfill material may be compacted in a single
lift. Judging the amount of water required and experimenting
with the thickness of the lifts is largely a trial-and-error process.
You'll probably have to test repeatedly to get progress reports
on the success of the different methods and techniques. In
difficult cases, it might be worth the cost of employing the
services of an independent soil testing laboratory.

The same pipeline may go through damp soil (near optimum
moisture), dry soil, and overly wet soil. It's hard to establish a
pattern, so you can't just instruct that a given amount of water
be applied or that a given amount of compactive effort is
adequate.

Failing the density tests can be a nightmare. Protect against it. Work cautiously. If necessary, test the compaction method you plan to use before production work begins. Be sure you have the right combination of equipment and methods before doing the lion's share of the work. And then, if possible, wait for wet soil to dry to below optimum moisture levels before making the density measurements required in the specifications.

7

INSTALLING WATER SYSTEMS

As a nation, we can be proud of the coverage of our water systems. Virtually every town and city in the U.S. has a piped water system that provides water for drinking, cleaning, and fire protection. Replacing, extending and repairing these systems is the subject of this chapter.

Cities may draw their water from reservoirs, lakes or wells. Water not pure enough for human consumption is treated to make it safe, then distributed to the point of use through pressurized pipelines. When water is stored in a tank or reservoir that's higher than the distribution system, gravity creates pressure in the pipes. If the water source isn't higher than the distribution system, pumps are needed to create pressure so water will flow to the point of use.

The most efficient layout for a water distribution system is a *grid* as in Figure 7-1. Note that the lines are interconnected, allowing water to feed any point of use from several different directions. When there's a sudden high demand for water in one area, the entire system responds. Flow rates go up in all mains.

The grid layout is important when there's a fire, for example. A fire hydrant on a 6-inch dead-end (non-grid) pipeline can draw only what's available through the single line. A fire hydrant on a 6-inch pipeline in a water grid can draw water from two directions and from several mains.

Grid layouts also help equalize pressure throughout the system and minimize *water hammer*. Water hammer occurs when a sudden demand is placed on a pipeline. Water rushes into the pipeline to replace the water used, causing a water surge. This creates a vacuum. When the valve is closed, the momentum of rushing water hammers against the valve and increases pressure momentarily at points throughout the system.

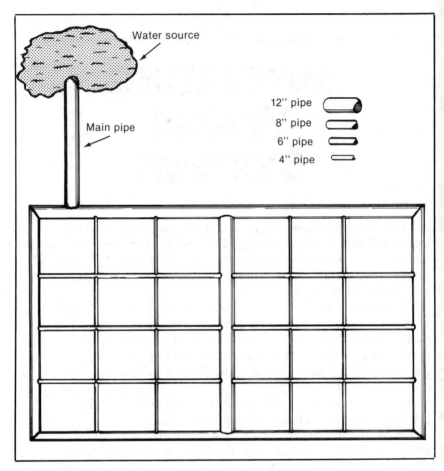

Simplified water grid
Figure 7-1

Surges and vacuums are hard to remedy and can destroy pipes and valves. A grid layout reduces water hammer by reducing surges and vacuums.

Pressure pipe and a grid layout are the key to every efficient water system. But there are other points to consider when designing any water system: the effect of water pressure and thrust on the system, pipe material used, fittings and joints selected. The efficiency of your equipment and crew and final testing of the completed work can also have an effect on system performance. Let's take a look at each of these key factors.

Water Pressure and Thrust

Every pipeline must be strong enough to withstand the maximum pressure of the water it carries. Water pressure is expressed as pounds per square inch, or *psi.* This means the water exerts a given number of pounds of force against each square inch of surface. Water pressures in municipal systems commonly range between 40 psi and 90 psi. Above 90 psi, the pressure is too great for most household plumbing. If the pressure is too high, a pressure-reducing valve must be installed in the pipeline to bring the pressure down to an acceptable level.

The three most common classes of water pipe are designed to hold water at 150 psi, 200 psi and 350 psi. A 150 class pipe, for example, can hold water up to 150 psi. But keep in mind that water pressure will vary with the demand, source of supply and layout of the system. In any gravity pressure system, for example, it's the *head* of the water that creates the pressure below the storage point. This means that the higher the supply point is above the point of use, the greater the pressure. Note that the system has to be designed so pipe at the lowest points can hold the most pressure.

Another factor that affects water pressure in the system is demand. Drawing water out of a pipeline will temporarily reduce the pressure. At night, when the demand goes down, pressure in the pipeline tends to increase.

The outward force of water pressure is called *thrust.* The effect of thrust against fittings, bends and dead-ends in a pipeline can be very powerful. Look at the thrust forces and estimated bearing load tables in Figure 7-2. For example, the thrust against a 12-inch 90-degree elbow at 100 psi is 16,000 pounds. To hold this elbow in place would require a thrust block of at least 32 square feet in soft clay, 16 square feet in sand, and 4 square feet in sand and gravel cemented with clay. Here's the formula for finding the size of the thrust block:

$$\frac{\text{Trust per 100 psi}}{\text{Bearing load of soil}} = \begin{array}{l} \text{required area of thrust block} \\ \text{per 100 psi of water pressure} \end{array}$$

Every pipe fitting you install has to be adequate to withstand the forces of water pressure and thrust. Proper installation of the pipe, fittings and joints is very important.

Pipe size in. (mm)	Fitting 90° elbow	Fitting 45° elbow	Valves, tees dead ends
4 (100 mm)	1,800 (8,007)	1,100 (4,893)	1,300 (5,783)
6 (150 mm)	4,000 (17,793)	2,300 (10,231)	2,900 (12,900)
8 (200 mm)	7,200 (32,027)	4,100 (18,238)	5,100 (22,686)
10 (250 mm)	11,200 (49,820)	6,300 (28,024)	7,900 (35,141)
12 (300 mm)	16,000 (71,172)	9,100 (40,479)	11,300 (50,265)

A   Thrust developed per 100 psi pressure, lb force (N)

Soil type	Lb/sq. ft.	N/m^2
Muck, peat, etc.	0	0
Soft clay	500	23,940
Sand	1,000	47,881
Sand and gravel	1,500	71,821
Sand and gravel with clay	2,000	95,761
Sand and gravel cemented with clay	4,000	191,523
Hard pan	5,000	239,403

B   Estimated bearing load of soil

Thrust forces and estimated bearing load of soil
Figure 7-2

Main Line Materials

Let's start with the main line in your water system. We'll look at the pipe material, fittings and joints you'll use for the main line. Then we'll go on to the water service pipe and look at the material, fittings and clamps required for service pipe installation.

Pipe Material

The water pipe must perform three important functions. It has to hold in the water pressure, withstand thrust, and resist the inward pressure caused by ground forces and vacuums that result from high demand or sudden changes in water pressure. For these reasons, water pipe is heavier and stronger than sewer pipe of comparable diameter. Here are six types of water pipe you should know about.

Wooden water pipe— At one time, wood pipe was used for water mains. Wood pipe was made the same way wood barrels

ıre made today. I doubt if you'll be laying any wood pipe, but
f you work in an older city, you may be called on to repair
,ome.

Ductile iron pipe— Available in the high 350-psi rating, ductile
ron pipe is too heavy to lay by hand. It's often specified in
ıreas where settlement is expected, such as trench crossings,
)ecause of its high load-bearing capacity. It's a very strong pipe.

Transite pipe— Transite pipe is also called A.C. (asbestos
:ement) pipe. It's becoming less popular because of the asbestos
)roblems. And smaller diameters have a reputation for
)rittleness. But it's inexpensive. The most common length is
.2'6". Used with fiberglass collars (bells), it's a good pipe
naterial. Transite pipe over 8 inches is too heavy to place by
ıand. It can be *stab assembled* if the inspector allows it. While
,uspended from a sling, the new length is swung into the bell of
he previously laid length. Two pipe layers swing the new length
)f pipe away from the bell of the previously laid length. As the
)ipe swings back toward the bell, it's guided into position in the
)ell. This saves time because each length doesn't have to be
)ried into place with a bar. But it also increases the chance of a
damaged bell.

PVC pipe— PVC pipe is comparatively light and easy to cut
ınd bevel. But high-pressure PVC pipe larger than 8 inches in
diameter is heavy and requires machine handling. PVC is
lexible and forgiving of abuse during warm-weather
nstallation. But it loses its flexibility during cold weather. Some
,mall-diameter (3 inches or less) PVC is fragile, so engineers
)ften specify full-cover bedding if the soil is rocky.

   Because PVC pipe is derived from petroleum products, its
)rice varies with oil prices. When oil prices are low, it's very
competitively priced.

   You can join PVC pipe with slip joint or glue joint gaskets,
)r compression couplers. The glue joining is most common for
,mall diameter pipe. One disadvantage of the glue joint is that
he surfaces to be joined must be perfectly dry. That's not easy
f you're working in wet laying conditions.

   Here's something to watch out for in joining PVC pipe. Some
'askets are designed to be used only one way. It's possible to
nsert the gasket into the bell backward. If you assemble the
)ipe with the gasket in backward, you may *fishmouth* the

gasket, pushing it out of the gasket seat during assembly. With some loose gasket systems, it's important not to lubricate the bell. If lubricant gets behind and underneath the gasket, it can cause the gasket to *fishmouth*, slip out of place during assembly.

Some cities outlaw PVC pipe in buildings because it gives off noxious fumes when it burns. This may limit its use to below-ground applications in some cases.

Fiberglass pipe— A key advantage of fiberglass is that it's both lightweight and strong. However, cutting, joining and repairing are all more difficult than with some of the more common pipe materials.

Cast-iron and steel pipe— Both have the disadvantage of corrosion. And cast-iron pipe is brittle. But modern coatings reduce the chance of corrosion and make them a good choice for some conditions.

Fittings

Pipeline fittings are used to change the pipe size, pipe type or water direction. Main line fittings include elbows (ells), crosses, tees, plugs and caps. These are illustrated in Figure 7-3. Notice that each fitting requires a thrust block, a mass of concrete placed to resist the thrust of water that is inevitable when water changes direction.

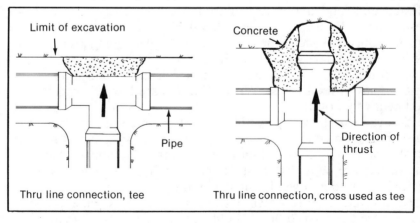

Line fittings and thrust blocks
Figure 7-3

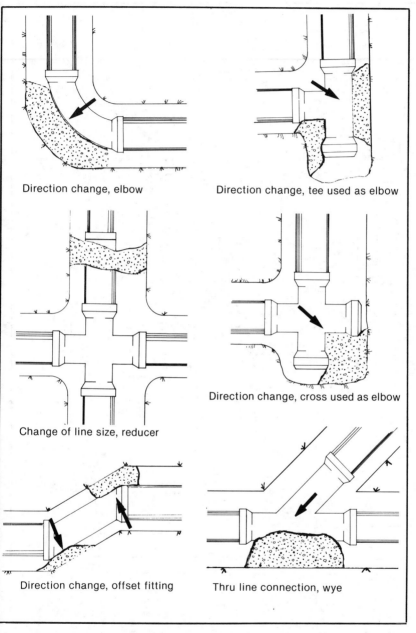

Direction change, elbow

Direction change, tee used as elbow

Change of line size, reducer

Direction change, cross used as elbow

Direction change, offset fitting

Thru line connection, wye

Line fittings and thrust blocks (continued)
Figure 7-3

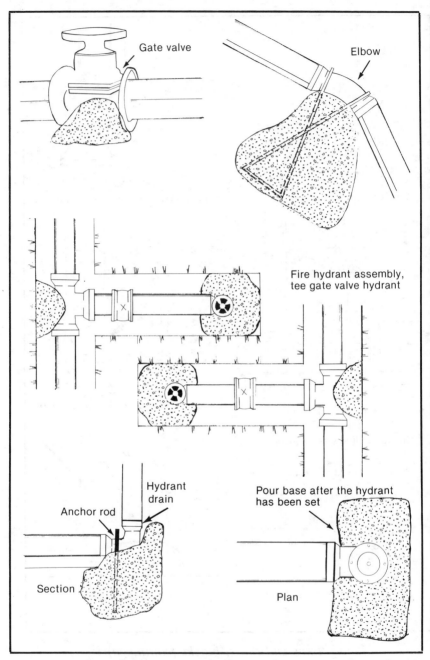

Gate valve

Elbow

Fire hydrant assembly,
tee gate valve hydrant

Anchor rod

Hydrant drain

Section

Pour base after the hydrant
has been set

Plan

Line fittings and thrust blocks (continued)
Figure 7-3

There are also three important *control* fittings that you'll use to regulate the flow of water in the main:

Gate valve (GV)— Opens or shuts off the flow of water through the main line. Some typical gate valves are shown in Figure 7-4.

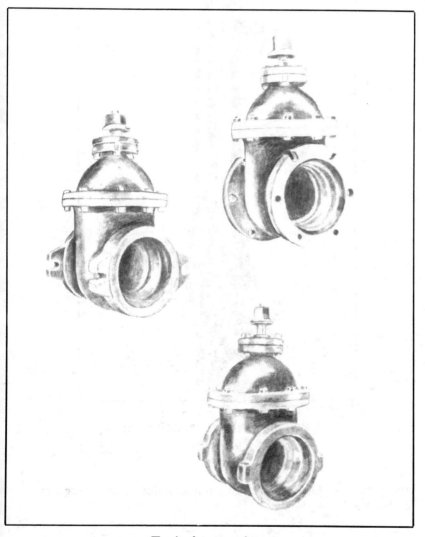

Typical gate valves
Figure 7-4

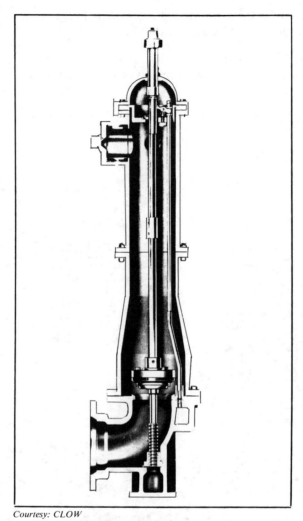

Courtesy: CLOW

Ground line hydrant in open position
Figure 7-5

Fire hydrant— Delivers a large volume water in an emergency.
See Figure 7-5.

Air relief valve— Allows air to escape from the main line when
there is no other escape. An air bubble in a pipeline reduces the
pipe's carrying capacity and contributes to water hammer.

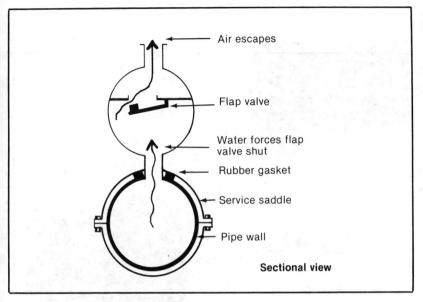

Air escapes

Flap valve

Water forces flap
valve shut

Rubber gasket

Service saddle

Pipe wall

Sectional view

Air relief valve
Figure 7-6

Avoid air bubbles in your line by installing it with constant, gradual slopes. Bubbles develop when air is trapped at a high point that isn't a point of use. If a high point in the line is unavoidable, be sure to install an air relief valve so that the air has a means of escape. A typical air relief valve is shown in Figure 7-6.

Joints
Three types of joints are used for main pipelines and fittings: slip or push joints, mechanical joints and flanged joints.

Slip or push joint— This is the easiest to install. It's used for almost all main line pipe connections. Slip joints are specified by Ringtite or other brand name. Figure 7-7 A shows a typical push joint. The push joint has one major disadvantage: thrust may cause the joint to come apart. As long as pipe is laid in a straight line, this isn't a problem. But any change of direction will need reinforcement. Joints can be reinforced by blocking

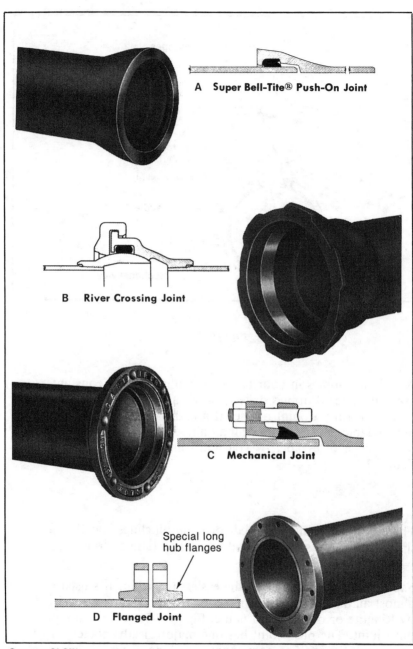

A   **Super Bell-Tite® Push-On Joint**

B   **River Crossing Joint**

C   **Mechanical Joint**

Special long
hub flanges

D   **Flanged Joint**

Courtesy: CLOW

Typical joint fittings
Figure 7-7

and by using specially cast fittings that allow the assembly to be tied together with bolts or threaded stock. Note the Clow river-crossing joint in Figure 7-7 B.

Mechanical joint— This is a compression coupling. It seals by mechanically forcing a wedge-shaped gasket against the outside diameter (O.D.) of the pipe and the inside edge of the fitting. This is done either by using bolts to pull in a follower or, in the case of small-diameter service fittings, by the force of the nut being threaded onto the fitting. Mechanical joints are less likely to come apart than slip joints, but they still require blocking. They are abbreviated MJ. See Figure 7-7 C.

Flanged joint— Is completely rigid after assembly. You'll use flanged joints for above ground pipe work, such as pump houses, or where a line must hold pressure immediately after assembly. Flanged fittings are sometimes used for valves, tees and crosses. Assemble the flanges above ground whenever possible. It takes time and room to clean both flanges, fit a gasket between them and tighten the numerous nuts and bolts. The abbreviation for flanged joints is Flg.

Fittings are available in each of the joint types, and a combination of joint types. If you see the notation MJ x Flg. G.V., it means a gate valve with one side flanged and the other side a mechanical joint.

Service Line Materials

We've looked at the pipe material, fittings and joints for water mains. Near the point of use, service lines connect with the main. Now let's see how pipe material, fittings and clamps are selected for service lines.

Pipe Material

Four materials commonly used for water service lines are polyethylene, galvanized steel, copper and polybutylene.

Polyethylene (PE)— Is flexible and easy to work with. You can use compression couplers, reinforced with steel inserts, to join PE pipe. Because of its flexibility, PE can be placed under ground with a plow rig, rather than being laid in a trench. But it's so soft that it can be cut and abraded by rocky soil. The engineer on your job will specify material to meet soil conditions. Be sure the specs allow for PE service pipe before you install it.

Galvanized steel— Was once commonly used for water service pipe. But some soils and chemicals in water corrode even galvanized steel. In suitable conditions, galvanized steel can last a long time. It's tougher than other pipe, so it's a good choice in areas where more pipe will be laid. Galvanized steel usually requires threaded joints. And you may need to cut and thread short lengths on the job. This will add manhours to your assembly time. Because galvanized steel is rigid, it's commonly used as the pipe that joins the water meter.

Copper— Is widely used for service pipe. Worries, real or imagined, of gophers chewing soft "poly" pipes are one of the reasons builders prefer copper. There are several ways to join copper pipe. Modern compression fittings make for fast, easy coupling. Another method is the flare joint. It's important to have two perfect mating surfaces so your joint will be watertight. Joints can also be soldered, but this is difficult under site conditions.

Polybutylene— Can be used in some situations to replace copper, and has similar characteristics.

Fittings
The six common types of service line fittings are threaded, compression, glue-joint, control, flanged and adaptor fittings.

Threaded fittings— These connections are often used for galvanized iron and PVC pipe. Pipe can be threaded on site, using a pipe vice and die. Pipe purchased in standard lengths will arrive already threaded. Short lengths of threaded pipe are called *nipples*. The spigot end of threaded pipe is always referred to as the *male* end. The coupler or nut thread is the *female* end.

All threaded connections require a sealing tape or paste compound. But don't use thread-sealing paste, or *pipe dope*, on plastic threads. You can use Teflon tape for both plastic and steel threads. Be sure to wrap the tape in a clockwise direction, so it doesn't unravel as the thread is tightened.

Compression fittings— Can be used on all types of pipe. But be sure to reinforce the relatively soft PE pipe with steel inserts. Compression fittings compress a gasket between two surfaces.

Glue-joints— Are used to connect PVC pipe. They're simple to assemble. Just be sure that the mating surfaces are dry when you apply the glue. The PVC will take the glue better if you use a primer to soften the PVC first. It's good practice to twist the pipe as you push the end together. This spreads the adhesive evenly.

Control fittings— Are needed to interrupt flow, reduce water pressure, and to shut off the water to a service line when required. Three important control fittings are the corporation stop, the stop and drain and the curb stop.

• Corporation stop: The primary control is the corporation stop, or *corp*. See Figure 7-8. This fitting is either threaded directly into the main pipe or threaded into a *service saddle* which is strapped to the main pipe. See Figure 7-9.

The corp operates on a different principle than a gate valve. The corp uses a tapped, drilled shaft enclosed in a tight-fitting body. Turning the shaft 90 degrees closes the hole and prevents water from passing through. Corps are made from bronze or gunmetal. These materials can be damaged by the unprotected jaws of a pipe wrench. So be careful during installation. There's a nut and washer at the bottom end of the tapered shaft that tighten the shaft into place. Old corps on service lines may have loose nuts. Tightening these nuts can help stop a leak. It's also a good practice to check the tension of the nuts on new corporation stops.

• Stop and drain: A more elaborate corp. The stop and drain interrupts the water supply and lets the water drain back from the pipe beyond the fittings. In cold weather, this lets you drain the pipes so they won't freeze. Be sure to install these fittings according to the direction arrows marked on the outside of the fitting. See Figures 7-10 and 7-11.

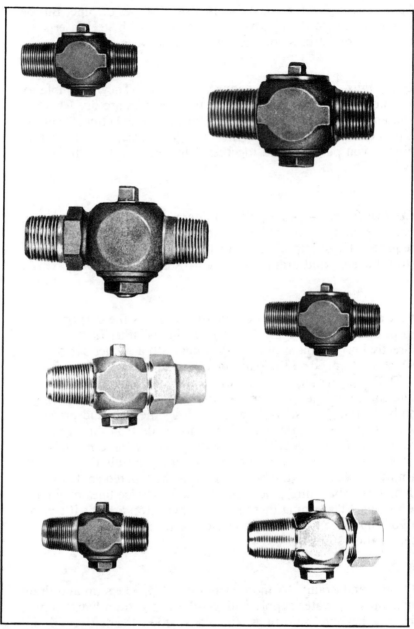

Typical corporation stops
Figure 7-8

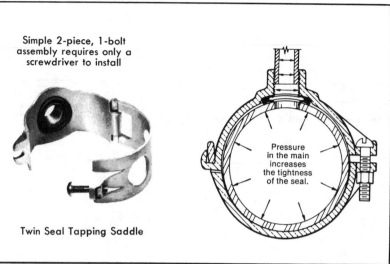

Simple 2-piece, 1-bolt assembly requires only a screwdriver to install

Twin Seal Tapping Saddle

Pressure in the main increases the tightness of the seal.

Courtesy: CLOW

Service saddle or tapping saddle for PVC pipe
Figure 7-9

Courtesy: A.Y. McDonald Mfg. Co.
Stop and drain
Figure 7-10

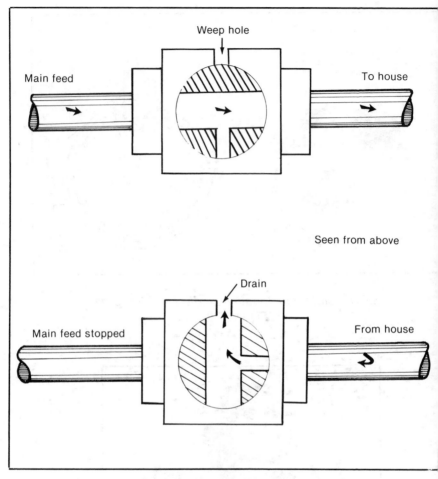

How a stop and drain works
Figure 7-11

• Curb stop: An in-line valve that works on the same principle as the corp stop. A typical curb stop is shown in Figure 7-12. Some service line specs demand both a curb stop and a valve in the meter box. The valve in the meter box is usually part of the meter setter, as shown in Figure 7-13. If there's a valve already provided on the meter setter, the specs will usually call for a corp stop instead of a curb stop. Typical water service line plans using corp and curb stops are shown in Figure 7-14.

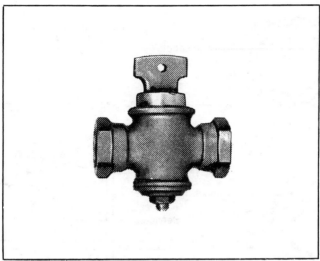

Courtesy: A.Y. McDonald Mfg. Co.

Curb stop
Figure 7-12

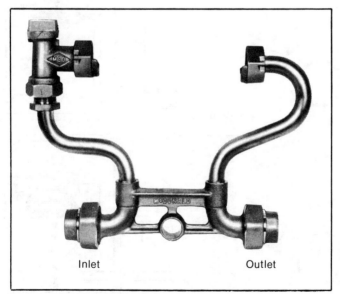

Inlet   Outlet

Courtesy: A.Y. McDonald Mfg. Co.

Meter setter with valve
Figure 7-13

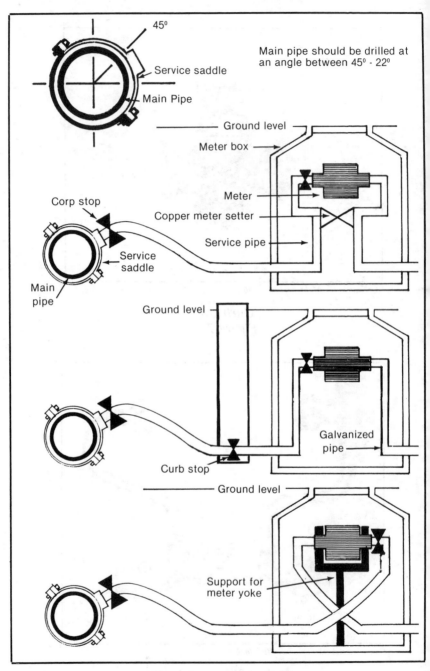

Typical water service plans
Figure 7-14

Flanged fittings— Used to hook up water meters so the meter can be lifted directly out of its setting. Meters with quick-detach flanged couplers measure the water used, either in gallons or cubic feet.

If you install a meter backwards, it will either work in reverse or it won't work at all. This causes confused meter readers and happy water consumers. Make sure you install the meter the right way.

Adaptor fittings— These allow you to switch from one type of fitting to another.

Repair Clamps

Pipe repair clamps are used to fix accidental damage or to correct problems created by sloppy installation. When using repair clamps on existing pipe, be sure the pipe is clean and free from scale. All clamps are equipped with high-tensile bolts and require adequate torque to ensure a watertight seal. And remember to torque the bolts down alternately. If you overtighten on one or both sides of a coupler, you'll end up with broken bolts and a clamp that's out of alignment.

Three commonly-used repair clamps are the compression coupler, the bandage-type coupler, and the split-ring bell repair clamp. Before we discuss the function of each type of clamp, take a look at Figure 7-15. It shows pipe outside diameters.

Size (inches) inside diameter	Cast and ductile iron (O.D.)	Spigot or MOA* Class 150 (O.D.)	Spigot or MOA* Class 200 (O.D.)	Pipe Class 150 (O.D.)	Pipe Class 200 (O.D.)	PVC (O.D.)
4	4.80	4.81	4.81	5.27	5.57	4.500
6	6.90	6.91	6.91	7.37	7.56	6.625
8	9.05	9.11	9.11	8.57	9.74	8.625
10	11.10	11.66	11.66	12.12	12.12	10.750
12	13.20	13.92	13.92	14.38	14.38	12.750
14	15.30	16.22	16.22	16.73	16.88	--
16	17.40	18.46	18.46	18.97	19.19	--

*MOA = milled overall

Comparative pipe O.D.
Figure 7-15

Courtesy: CLOW

Compression coupler
Figure 7-16

You'll need this information to select the right gasket and repair clamp.

Note the reference to MOA in Figure 7-15. MOA stands for "milled overall." You'll use MOA pipe for the short sections required in fitting assemblies. MOA pipe isn't the same as MEE pipe. MEE means "milled each end." MEE pipe is full-size pipe that has the middle section left in its rough-cast condition.

Figure 7-15 also shows that the spigot O.D. (outside diameter) is less than the overall O.D. of the full-size pipe. So joints made to the overall O.D. require compression gaskets that are larger than the spigot dimensions. Rough-cast middle sections of full-size pipe won't fit the bell or collar of the pipe. So compression couplers must be used.

Compression coupler— When a section of pipeline is damaged, you can cut out the damaged piece and clamp a replacement length into place with compression couplers. Figure 7-16 shows a typical compression coupler.

Bandage-type coupler— If the pipe has a break along its length, you can repair it using the wraparound, bandage-type coupler shown in Figure 7-17. You can also use a similar clamp, the saddle clamp, to repair a corp that has been torn out. See Figure 7-18.

Bandage-type repair coupler
Figure 7-17

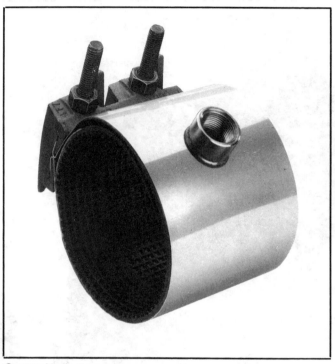

Courtesy: Mueller Co.

Saddle type repair clamp
Figure 7-18

Split-ring repair clamp— To repair a fishmouthed gasket, use the split-ring bell repair clamp shown in Figure 7-19. It's less expensive than cutting out a section of pipe and adding two straight couplers. Careful installation will prevent fishmouthed gaskets.

All of the materials you use in your water system work require careful inventory and storage. Treat these materials with the respect they deserve. Have some organized system of storage that prevents damage and loss until the material is needed for use. Many of the parts are small and easy to misplace. Missing parts can cause delay, inconvenience and inflate labor costs. Make sure your crew has easy access to the materials they need.

We've discussed the key materials required for installing the main pipeline and the service lines. Next you'll need to select the right equipment and crew to do the job.

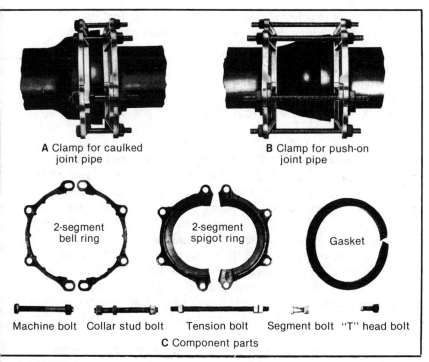

A Clamp for caulked joint pipe

B Clamp for push-on joint pipe

2-segment bell ring

2-segment spigot ring

Gasket

Machine bolt   Collar stud bolt   Tension bolt   Segment bolt   "T" head bolt

C Component parts

Courtesy: CLOW

Split-ring bell repair clamps
Figure 7-19

Selecting Equipment

Water pipe is installed relatively shallow, just below the frost line. It may be as shallow as two feet in some areas. Four to five feet of cover is adequate for most climates. In extremely cold climates, you may have to go down six or seven feet and also insulate the pipe. A small or midsize wheeled backhoe can easily handle a trench five or six feet deep. The versatility of this machine makes it the most popular choice for water line installation.

Another choice is the smaller tracked excavator. This is useful when the pipe has to be bedded and covered with imported fill material. Trenching machines excavate quickly when there aren't any obstructions. But they have a hard time maneuvering around utility cross-lines, so they're not practical where cross-lines are present.

On projects involving more than 5,000 linear feet, you'll probably want to use at least one loader backhoe. In some cases, you'll also use additional backfill and compaction equipment.

Production schedules vary with site conditions. A wheeled backhoe trenching constantly in a developed street can handle about 100 linear feet of 4-foot deep trench per hour. Traffic, asphalt removal and numerous cross-lines may cut the rate by 50%.

Make sure your backfill and compaction equipment can handle the trenching schedule. An excavation rate of more than a few hundred feet a day requires high-capacity compaction equipment. You may also need hand-held compaction tools to compact the smaller areas. A hoe-mounted vibratory compactor is a good choice for compaction in small areas.

If you need to trench over 1,000 feet a day, use a wheeled loader to do the backfilling. And get a loader equipped with a quick-detach front bucket that can also be fitted with forks to handle pipes and pallets of fittings. These attachments double the utility of your loader, letting it handle both the backfilling and material layout.

Some water pipe material is extremely heavy and requires machine laying. Figure 7-20 shows the weights of commonly-used pipe materials. You may be able to use your trenching equipment to lay the pipe. But it may be more productive to bring in specialized equipment to do the laying. A sideboom pipe layer is ideal for laying pipe. See Figure 7-21. The versatile backhoe loader is often used for this task as well. A pipe-laying hoe can also handle trench backfill.

		Lbs. per lin. ft.
Asbestos cement pipe	Class 150	18.7
PVC pipe	Class 150	9.00
Fiberglass composite pipe	Class 350	4.55
Ductile iron pipe	Class 350	24.40
Reinforced plastic mortar pipe	Class 150	6.00

Typical pipe weights
Figure 7-20

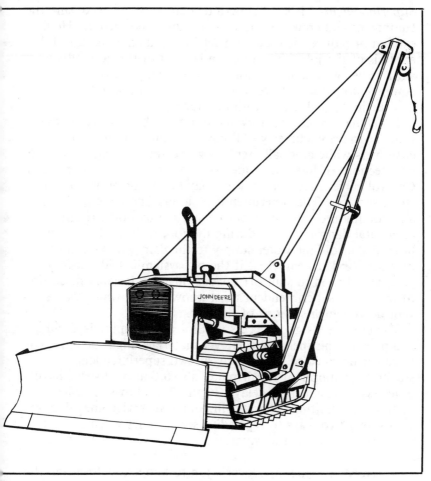

Courtesy: John Deere

John Deere dozer equipped with a sideboom
Figure 7-21

Forming a Crew

The minimum crew required for installing a water system
includes a superintendent, an equipment operator, a pipe layer
and a pipe layer's helper. To increase the production rate, hire
additional laborers to assist with material layout, backfill
compaction and damage repair. Let's look at the responsibilities
of each of these crew members.

Superintendent— It's the superintendent's job to make sure the equipment and crew can meet the production schedule. He'll expand or reduce the crew to meet job requirements. Careful analysis of the job site is an essential part of this job. When work is being done in developed areas among existing lines, equipment operators have to be able to work around obstructions quickly and with accuracy.

Installing a replacement water system is the most complex water work. Sometimes you'll have to maintain service to customers while gradually replacing and transferring services to new sections of line. Old water systems are often in poor repair. Control valves may not work or may be lost or buried under street surfaces. The superintendent should know the position and condition of all valves before starting to work. If any accidental damage occurs during the job, he'll need to know how to shut off the water supply to the damaged section of pipeline. And remember — if the old system is laid out in a grid, you'll need to shut off *several* valves to isolate one section of pipe. The superintendent must know exactly which valves shut down which sections of the line.

The crew needs good access to all necessary materials and equipment. This includes whatever materials might be required to repair unexpected damage. A shrewd superintendent stockpiles a supply of repair materials so that he doesn't have unnecessary production delays. Figure 7-22 shows several methods of shutting off water in damaged service lines. Figure 7-23 shows two ways to repair live service lines when it's not practical to shut off the water.

Pipe layer— It's the pipe layer's job to know how to install the parts and fittings quickly and efficiently. Pipe weighing over 100 pounds per joint (section) is usually laid with lifting equipment. It's essential that the pipe layer measure accurately when cutting the pipe. Fittings, such as control valves, fire hydrants, tees and crosses, must fit onto the pipe perfectly at the specified interval.

The pipe layer can boost the production rate by measuring and preparing the pipe and fittings for installation before the trench is excavated. Here's an example. Let's say the crew is installing a cross-assembly consisting of the cross, two short lengths of pipe and two gate valves. The pipe stubs can be cut and prepared and the gate valves fitted to the cross before the trenching hoe reaches the intersection. When the hoe does reach

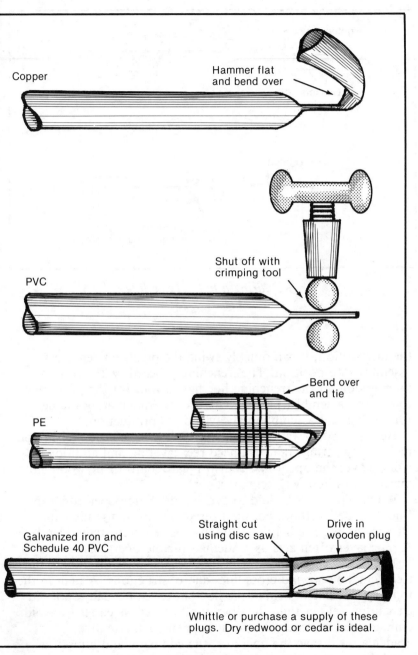

Repairing damaged service lines
Figure 7-22

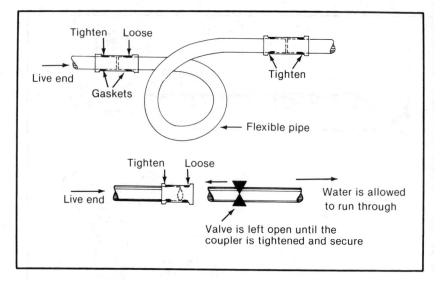

Repairing a live service line
Figure 7-23

the intersection, it can quickly swing the preassembled cross-assembly into position. The trenching proceeds without interruption. If the trenching hoe has to wait for the pipe crew to cut and assemble, productivity drops. Anticipating the need for cutting and assembling always boosts production.

Here's another way your pipe layers can save time. When you have to lay fittings that will need reexcavating and additional work, cover the open pipe end with cardboard or plywood and mark its position with scrap 2 x 4's.

Water pipe must be laid to even grade. Some operators can "hold grade" without constant directions from a grademan. Others will require help to maintain the required trench depth. A pipe layer often acts as grademan for the operator.

The pipe laying crew may also be responsible for guiding the trenching hoe around cross-lines and obstructions. A man in the trench immediately behind the hoe bucket can usually spot old trench lines easier than the operator. It's always easier to avoid damage than to repair it after it happens. An attentive grademan will save the operator, the company and himself a good deal of trouble and expense by helping the operator avoid damage.

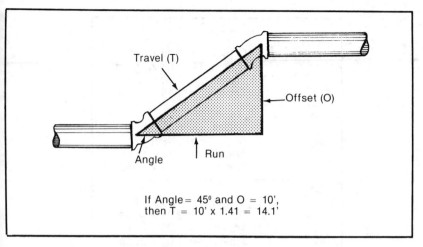

If Angle = 45° and O = 10',
then T = 10' x 1.41 = 14.1'

Calculating the length of an offset
Figure 7-24

The pipe layer often has to calculate the precise position of
trenches and pipelines for an *offset*. An offset is a combination
of elbows or bends that brings one section of pipe into line
with another section of pipe. See Figure 7-24. Notice that the
displaced section of pipe remains parallel to the original
pipeline. The angle of the elbows determines the *factor* you'll
use for finding two other important dimensions: *run* and *travel*.
Figure 7-25 shows what these terms mean and gives you the
factors for common elbow angles.

Here's how to use the factors in Figure 7-25. Let's say we're
using two 60-degree elbows to align two pipelines. The pipelines
are parallel and 30 feet apart, so the offset is 30 feet. Figure
7-25 shows that, for 60-degree elbows, the run equals 0.57 times
the offset. Here's the calculation:

$$30' \times .57 = 17.1' \text{ run}$$

Let's look at another example. If the elbow angle is 45
degrees and the offset is 10 feet, here's how to find the travel:

$$10' \times 1.41 = 14.1' \text{ travel}$$

60⁰

Offset	=	1.73 x run
Run	=	.57 x offset
Travel	=	2.00 x run
Travel	=	1.16 x offset

45⁰

Offset	=	run
Run	=	offset
Travel	=	1.41 x run
Travel	=	1.41 x offset

22½⁰

Offset	=	.41 x run
Run	=	2.41 x offset
Travel	=	1.08 x run
Travel	=	2.61 x offset

11¼⁰

Offset	=	.20 x run
Run	=	5.03 x offset
Travel	=	1.02 x run
Travel	=	5.13 x offset

Note: These figures are correct to two decimal places or have been rounded down to the second decimal place.

Factors for common elbow angles
Figure 7-25

Once the pipe crew has this information, they can mark the trench line for the excavating hoe.

If the water must be turned on immediately after the offset is installed, partially open the feed valve. This will help reduce damage if the assembly blows apart under pressure. Reducing the water feed won't reduce the thrust, but it will reduce the amount of water that escapes if a mishap occurs. It's important to brace the fitting against thrust immediately. You can do this with timbers and by pouring "high early" concrete. If at all possible, avoid putting immediate stress on the assembly.

We've discussed the equipment and crew we'll need for our water system work. Specifications usually demand that the new main line be tested and disinfected before hooking up the services. Before we examine how to run the final tests on our completed lines, here are some important tips to remember when you're installing the service lines.

Installing the Service Lines

Once the main line is in place, installing the service lines comes next. Domestic service usually requires 3/4-inch to 1-inch diameter pipe. Businesses may require 2-inch to 6-inch service lines. Look again at the typical water service plans shown back in Figure 7-14.

Joining the Main Line

The two methods of fitting domestic service lines into the main pipeline are the direct tap and the saddle tap. When you tap the main line, you'll also install the control fitting. Here's how.

Direct tap— Use the direct tap where the main line walls are thick enough to allow drilling and tapping. You can thread the corporation stop directly into the wall of the main pipe. The Mueller Company manufactures a direct tapping machine specifically for this task. See Figure 7-26. This machine allows you to install corp stops when water in the main pipe is under pressure.

Saddle tap— If the main line walls aren't thick enough to allow drilling and tapping, use the saddle tap. Figure 7-27 shows a saddle tapping machine. This also allows corp stops to be installed when the main is under pressure.

In some cases you can install the saddle while installing the main line. Use an electric drill to make the hole in the main pipe. This method is usually practical only in undeveloped subdivisions where the main pipe can be left exposed and the service trench is excavated away from the main line.

When installing service lines in developed areas, you'll probably be trenching from the property line toward the main pipeline. If you install your saddle before installing the main line, it'll be difficult to locate the saddle after it's buried. And it's hard for a backhoe to cross an open trench. It's easy to knock off the saddle and protruding corporation stop. Stay away from prior saddle installations when you're working in developed areas.

Control fittings— Specifications often require that you install the corp at an angle between 22½ degrees and 15 degrees off horizontal. This allows for some slack in the service pipe if trench settlement puts tension on the service line.

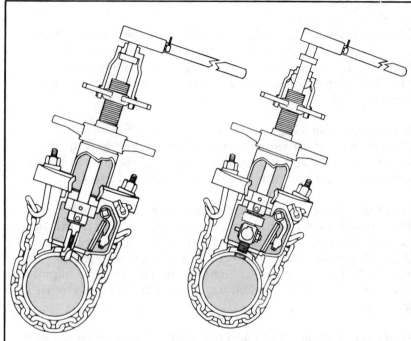

Mueller Machine drills and taps a hole in a pressurized main without water escaping.

The drilling and tapping tool is extracted and a corporation stop placed on the boring bar for insertion into the main.

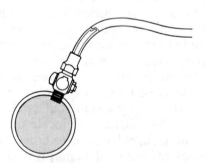

The closed corporation stop is now installed, ready for the service pipe to be connected.

The connection is now complete and the corporation stop opened

Courtesy: Mueller Co.

Direct tapping machine
Figure 7-26

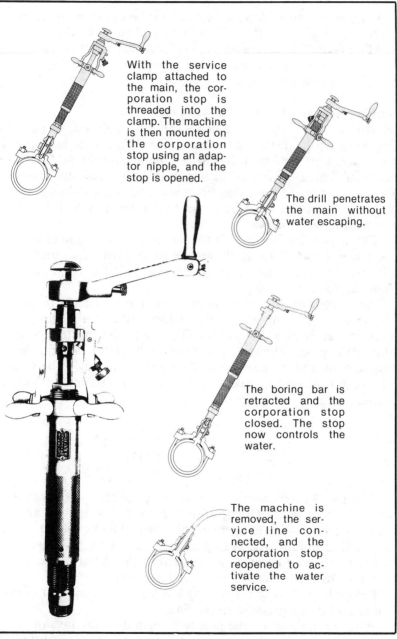

With the service clamp attached to the main, the corporation stop is threaded into the clamp. The machine is then mounted on the corporation stop using an adaptor nipple, and the stop is opened.

The drill penetrates the main without water escaping.

The boring bar is retracted and the corporation stop closed. The stop now controls the water.

The machine is removed, the service line connected, and the corporation stop reopened to activate the water service.

ourtesy: Mueller Co.

Saddle tapping machine
Figure 7-27

It's essential that *all* plumbing fittings be tightened securely. Every slight weep or drip will get worse. And if the line is holding relatively high pressure, escaping water can actually wear away parts of a metal fitting. Remember, it's easier to do it right the first time than to repair it later.

Joining New and Existing Service Lines

A *tie-in* is a joint that ties a new section of main line to an existing section. Installing tie-ins requires meticulous planning. Once you start a tie-in, it has to be completed promptly because the water has to be cut off during installation. In a business district, you may have to do tie-in work during non-business hours. Always give advance notice to consumers who'll be temporarily without water.

The most common tie-in is the straight compression coupler we saw in Figure 7-16. It allows you to join two pipes made of different materials. Be sure to select the correct gasket and follower for the new pipe.

Pipes made of different materials will have different outside diameters. For example, there's 0.8-inch difference in the outside diameters of Johns Manville Class 200 transite pipe and their PVC pipe. Whenever you're in doubt about the outside diameter of a pipe, measure it. That's the only way to be sure you've got the correct gasket. Here's the formula for finding the outside diameter of a pipe:

$$\text{Diameter} \quad = \quad \frac{\text{circumference}}{3.1416}$$

Don't make the common mistake of excavating too small a hole around a tie-in. Cramped quarters will slow the job and increase labor costs. Remember that thrust blocks will be required to hold the tie-in in place after assembly. But also be sure never to overexcavate in the area of the thrust blocks. Figure 7-28 shows how to excavate for a tie-in.

Before beginning the tie-in, lay out your excavation carefully. Then shut down pressure to the line.

When cutting starts, the pipe will begin draining into the excavation. You'll have to pump it out. When you're cutting into a line that has several hundred feet of large-diameter pipe, there's a lot of water in there. Have enough pumping capacity

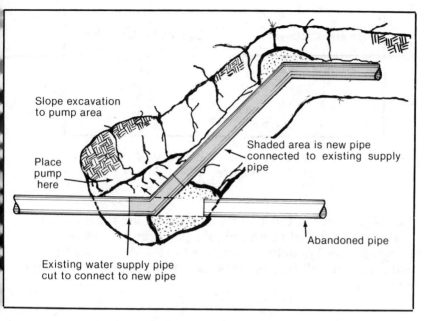

Excavating for a tie-in
Figure 7-28

to handle the water you need to remove. Figure 7-29 shows the
water capacity, in gallons per linear foot, of the common pipe
sizes. Here's the formula for calculating pipe capacity if the
answer you need isn't in the table.

$$\frac{\text{Diameter}^2 \text{ x } .7854}{144} = \text{cu. ft. capacity}$$

To convert to gallons:

Cu. ft. capacity x 7.48 = U. S. gallon capacity per linear ft.

When cutting into a water-filled line, use the method shown
in Figure 7-30. This will keep water from being thrown toward
the operator. With the first cut, water is thrown up and away
from the operator. Now the water can drain out of the pipe.
With the second cut, the water is thrown down and away from
the operator. Finally, the top of the pipe is cut. The operator

Pipe size (inches)	Gallons per linear foot
15	9.17
12	5.87
10	4.07
8	2.61
6	1.46
4	.65

Water capacity per linear foot of pipe
Figure 7-29

will still get wet, but this technique will minimize the drenching. If there's a great deal of water in the pipe, cut a chock out of the pipe to speed draining. "Measure twice and cut once" is a good maxim for all pipe-laying work. This is especially true with tie-in work.

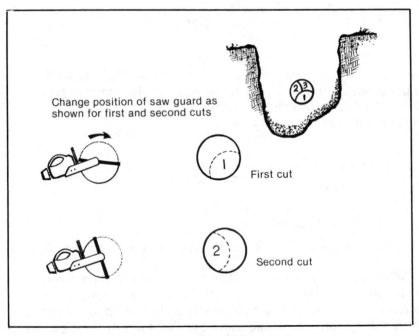

Change position of saw guard as shown for first and second cuts

First cut

Second cut

Cutting a water-filled pipe
Figure 7-30

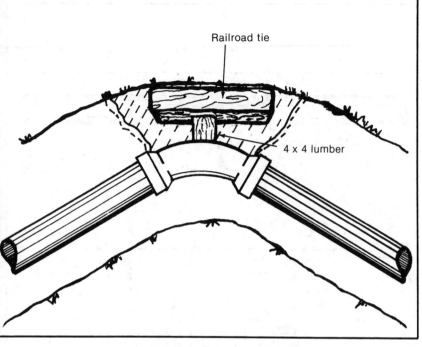

Railroad tie

4 x 4 lumber

Temporary bracing
Figure 7-31

Figure 7-2 listed the thrust force each 100 psi of water pressure places against pipeline fittings. For example, a 12-inch 45-degree elbow must endure 4½ tons of thrust. Thrust blocks are designed to resist that pressure. Concrete is the only acceptable thrust block. Use rapid-hardening concrete if the installation will be put under immediate pressure. While the concrete is hardening, support the tie-in with lengths of waste steel pipe about 4 inches to 8 inches in diameter. Stout lumber also makes an adequate temporary support. Figure 7-31 shows how a temporary bracing of railroad ties and 4 x 4 lumber is arranged before it's covered with rapid-hardening concrete.

Some tie-ins will be made with a tee. In Figure 7-32, the tee is fixed to the existing pipe with two straight compression couplers. If the pipe can be beveled for a slip joint or if you can use a mechanical joint, only one compression coupler will be needed. The drilling machine shown in Figure 7-33 is designed for work like this.

Plan view

Lines of cut

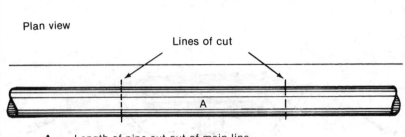

A   Length of pipe cut out of main line

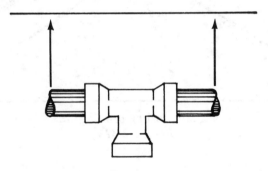

B   Tee fitted with pipe stubs ready for positioning

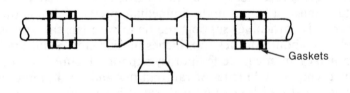

Gaskets

C   Tee fixed to existing pipe using straight compression couplers

Cutting in a tee
Figure 7-32

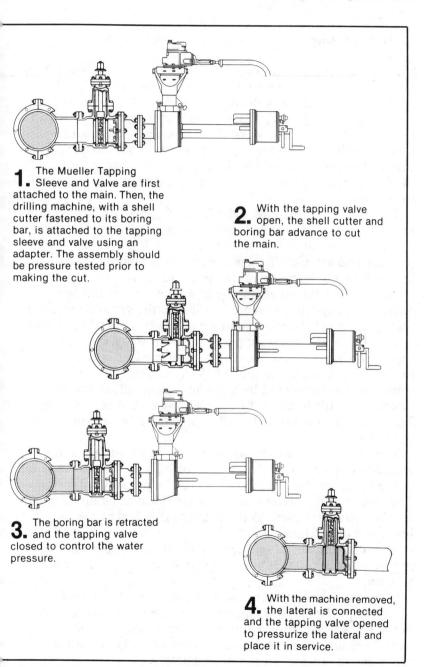

1. The Mueller Tapping Sleeve and Valve are first attached to the main. Then, the drilling machine, with a shell cutter fastened to its boring bar, is attached to the tapping sleeve and valve using an adapter. The assembly should be pressure tested prior to making the cut.

2. With the tapping valve open, the shell cutter and boring bar advance to cut the main.

3. The boring bar is retracted and the tapping valve closed to control the water pressure.

4. With the machine removed, the lateral is connected and the tapping valve opened to pressurize the lateral and place it in service.

Courtesy: Mueller Co.

Drilling machine
Figure 7-33

Testing the Line

It's standard practice to test each line under high pressure. You need to find out if the system will do what it was designed to do — hold water under any pressure condition. A system normally operating at 70 pounds per square inch may need to withstand pressures of double that when demand is low. The engineers should specify pipe and fittings that meet job requirements. For common operating pressures of 70 pounds per square inch, they'll probably specify materials designed for operation at 150 to 160 psi.

The strength of the pipe material isn't the only consideration. Fittings have to be installed so they won't come apart. Look again at the thrust blocks shown in Figure 7-3. The high-pressure test will identify leaks and cause a failure if one is likely under the highest working pressure likely.

You can test the pipelines one at a time, or you can test sections of a system together. Just isolate the section that you're testing. Shut off the main control valves and each of the meter setters. Use a high-pressure pump to pump water through a service tap into the test line. Forcing extra water into a line raises the pressure inside the line. The increased pressure is measured and monitored by a pressure gauge fitted into the feeding line. It's important to use a valve to isolate this gauge. The gauge can't withstand the full pressure of a working pump. See Figure 7-34.

The test pressure will break apart any faulty fittings and assemblies. To measure the amount of leakage, let the lines hold the test pressure for a period of time. The pressure will drop. Then measure the amount of water you have to pump back into the lines to return them to the pressure they had at the beginning of the timed period.

A certain amount of leakage is considered acceptable. The specification for your job will explain how much leakage is allowed.

When you're conducting the high-pressure test, remember that *excess air in the line will make your test results inaccurate.* Air in the line will be absorbed by water when you increase the water pressure. Water can absorb some air without noticeably increasing in volume. Here's an example.

If a cylinder has 50% air and 50% water under compression, the air will compress (not the water) as pressure is increased.

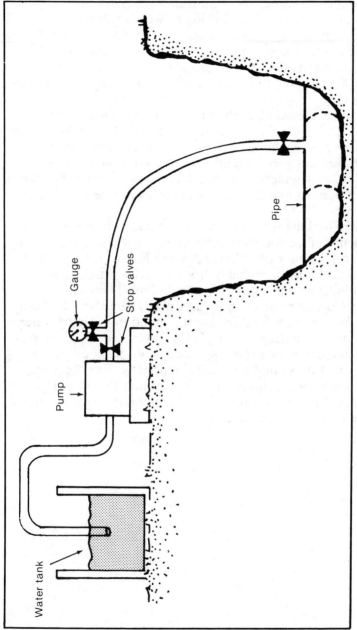

Pressure testing water lines
Figure 7-34

Under enough pressure, the compressed air will even "disappear" into the water, leaving us with a volume approximately the same as the volume of the water by itself. When we reduce the pressure, air will slowly escape from the water. This means that if your pipelines contain a significant amount of air during the high-pressure testing, your test results will be wrong.

Water lines shouldn't contain any air at all. That's why it's important to lay the lines to grade and to install air relief valves at any high points where air might be trapped. Even correctly installed, well-designed systems may contain some air after the first filling. Just remember to expel all air prior to testing. You may even need to install a tap for the sole purpose of releasing trapped air.

If you're looking for a leak, flush the line to make sure you expel all air before you start excavating. Then retest and try to establish the amount of leakage. A small leak of a few pints an hour indicates a drip, probably from a service fitting. Larger amounts indicate a problem in the main line. There are sophisticated electronic listening devices available now that can pinpoint the problem by detecting the sound of escaping water.

Be aware of the danger of exploding plugs and caps. Internal line pressure can blow off a plug or cap with terrific force. So stay clear of fittings under pressure, especially when lines are fitted in vaults or manholes. And be sure that the pressure has been released before you remove any temporary braces. A word to the wise should be sufficient.

8

INSTALLING
SEWER
SYSTEMS

The sewage volume produced in most communities is about equal to its water use — about 40 gallons per person per day. This effluent must be collected, transported and treated. As a utility contractor, you don't need to worry about treating the effluent. Your job is to install the pipelines that collect and transport it to the treatment plant.

In this chapter, I'll explain how to install the pipes to meet the design specifications. We'll discuss the materials, equipment and crew you'll need to do sewer system work. Then we'll go step-by-step through a sample sewer pipe installation.

Septic Tank Systems

Septic tank systems are used to hold and process sewage in rural areas and housing developments not served by municipal sewer systems. A septic tank is basically a settlement tank. The solids in the effluent settle to the bottom of the tank. The lighter fluids flow out of the tank and are released to underground pits and drains where they can be absorbed by the soil.

Effluent from a septic system isn't clean and potable. It's a definite health hazard if it filters down to an underground water table used for drinking water. That's not usually a problem in rural areas where homes are spaced well apart and most underground water is used only for agricultural purposes. But the quantity of sewage generated in towns and cities would poison underground water sources if the effluent weren't collected and treated properly.

Municipal Sewer Systems

Unlike the septic tank system, the municipal sewer system consists of laterals and mains that collect and transmit sewage to the point of treatment and eventual discharge.

Gravity flow is the cheapest and simplest method of transporting sewage. But gravity systems aren't always possible. Sometimes pumps and pressure pipe are needed to lift the effluent to a treatment plant. These are called *lift stations.*

Pressure pipe for sewer systems is laid about the same way as pressure pipe for water systems. It can be laid relatively shallow, but it must be laid to even falls and grades. Unlike pressure pipe, gravity lines must flow downhill regardless of the lay of the ground. Where gravity pipe has to run deeper than 30 feet, it may be necessary to tunnel or bore a hole for the lines. The alternative is to install lift stations and force mains (pressurized lines) for the sewage.

Whether the design uses gravity or pressure lines, every sewer pipe you install will have a specific carrying capacity or *rate of flow.* There are three factors that determine the rate of flow for a pipe: pipe diameter, slope, and pipe material.

Pipe diameter: Increasing the pipe diameter increases the carrying capacity. When you double the pipe diameter, the carrying capacity increases about *four times.* This means that an 8-inch diameter pipe has four times the carrying capacity or flow rate of a 4-inch diameter pipe.

Slope: Slope or grade in a pipeline causes the fluid to flow. The slope of a pipe may be as little as 0.01% (0.0001) in a large diameter pipeline. Small diameter pipe is generally laid to a steeper slope: 0.24% is about the minimum for 8-inch diameter PVC pipe. This 0.24% slope equals about 3 inches per 100 feet. That's a very gradual slope. It's just 2.4 hundredths of a foot per 10 linear feet. You remember that one hundredth of a foot equals about 1/8-inch, so in carpenter's measurements, 2.4 hundredths equal about 5/16-inch. That means the pipe will fall only 5/16-inch in 10 linear feet.

It takes great care to position the pipe so that it "falls" or slopes correctly. It's precision work. If it isn't properly done, the pipeline won't slope evenly. When a pipe is laid level or actually slopes the wrong way, it's called a *backfall.* Any backfall will be obvious to the inspector because water will pond in the low spots. Because backfalling pipe effectively reduces the usable diameter of the pipeline, specifications are often stringent. Many demand a grade (slope) tolerance of within 0.01 foot. It takes an accurate laser set-up (described in Chapter 3) and extremely precise pipe laying to meet this tolerance. Do it once and do it right!

Pipe Materials

The engineer may specify a particular class or standard of pipe. But sometimes you may use other pipe that meets the prescribed standard. This means that you, the contractor, can select the pipe material for the job. Three important factors to consider when selecting pipe material include: cost, strength and ease of installation.

Cost: Price will vary with demand and the cost of raw materials. For example, PVC pipe prices tend to go up when petroleum prices increase. And if there's a high demand for a certain type of pipe in your area, you can count on your local supplier raising his prices. Check current prices before you commit yourself to a specific type of pipe. Most pipe suppliers are eager to help you by providing price quotations.

Strength: Bedding conditions and trench width both affect the strength of the pipeline. Here's why.

Most sewer systems use the force of gravity to transport the sewage. A gravity sewer pipe doesn't have the water pressure and vacuums inside the pipe that stress pressure pipe, but it must be strong enough to resist the weight of the earth on top of it. Some pipe, such as PVC, is relatively soft and can flatten out (or squash) if a heavy weight is placed on it. More rigid pipe material such as concrete and clay resists flattening but is more brittle.

Pipe bedding can help strengthen either kind of pipe. Its purpose is to reinforce the pipe by providing firm support around its walls to resist the flattening effect and to provide a firm base along the length of the pipeline. Figures 8-1 and 8-2 illustrate the purpose of pipe bedding. If the natural ground is soft and likely to move, you need an additional form of bedding called foundation stabilization. See Figure 8-3. Crushed rock is often used under the pipe bedding to provide additional stabilization.

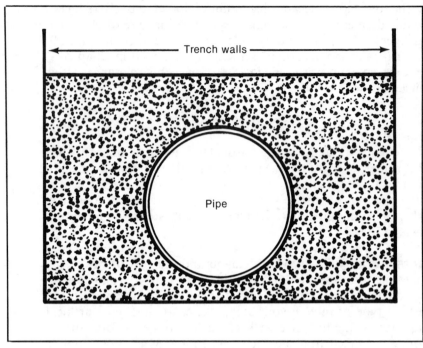

Compacted bedding provides support around pipe
Figure 8-1

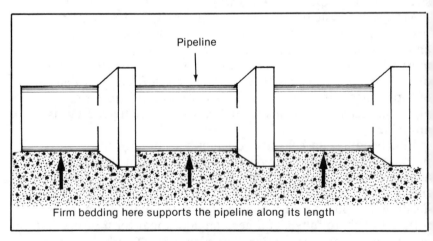

Bedding supports pipe along its length
Figure 8-2

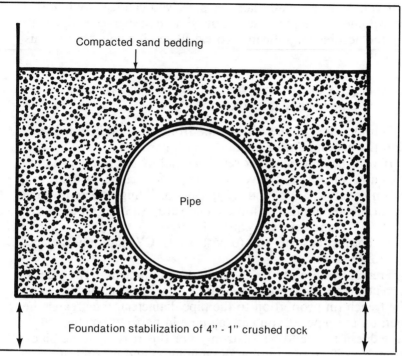

Foundation stabilization
Figure 8-3

A properly bedded line won't collapse under the anticipated load. But if you install the line without sufficient bedding, the stress of the weight of the backfill may twist and bend the line, causing it to leak or even collapse.

You can use bedding to "upgrade" a pipe material into another class. For example, if the engineer requires ductile iron pipe in a heavy-load situation, the specifications may state that only ductile iron can be used if no other pipe can meet the high-strength requirements. But you might be able to use a concrete encasement so another type of pipe meets the strength requirements.

Here's the formula for calculating bedding requirements:

Depth of bedding under pipe + diameter of pipe
+ bedding over pipe x trench width x length of
pipeline ÷ 27 = B. CY of bedding required.

If you're using small diameter pipe, you can usually ignore the volume of the pipe itself. But with large diameter pipe, this volume becomes significant. To calculate pipe volume, use this formula:

Diameter x diameter x .7854 x length of pipe (in feet) ÷ 27
= pipe volume in cubic yards

Subtract the volume of the pipe from the total bedding to find the net amount of bedding you'll need.

If your bedding supplier sells material by the ton instead of the cubic yard, you'll have to convert loose cubic yards to tons. A ton equals 2,000 pounds. Typically, bedding material weighs about 3,000 pounds per loose cubic yard. So one ton equals 0.67 L. CY and 1 L. CY equals 1.5 tons.

To convert cost per ton to cost per L. CY, multiply the cost per ton by 1.5.

Trench width also affects strength requirements. The sides of a trench resist downward pressure of the backfill. So the wider the trench (in proportion to the pipe diameter), the greater the load on the pipe. If you can lay pipe in narrow trenches, the pipe doesn't have to be as strong. For this reason, some pipe-laying specifications outline the minimum and maximum acceptable trench width.

Ease of installation: Some pipe is easier to install than others.

Let's keep these criteria in mind as we look at the five most common sewer pipe materials: concrete, clay, ductile iron, transite and PVC.

Concrete— Few materials are competitive with concrete in the large-diameter sizes (over 24 inches). Large-diameter reinforced concrete pipe has a high load-bearing capacity. The main disadvantage of concrete pipe in the smaller sizes (4-inch to 10-inch) is its heavy weight. In the smaller sizes it's available in 3-foot to 4-foot lengths. Tees or service taps are usually cast as part of the pipe. Difficulty in cutting concrete makes the position of the tee variable only within 2 feet.

Laying small-diameter concrete pipe requires firm bedding. This makes laying difficult in high groundwater. Assembling small-diameter concrete pipe and positioning the gasket is a fine art requiring skill and strength on the part of the pipe layer.

Clay— Like concrete, clay has a high load-bearing capacity. Clay pipe is lighter than concrete. It also has fixed tees and wyes. Joints used to be made with tarred rope and mortar joints. Now there are modern gaskets you can use with clay pipe that take less time to install.

Ductile iron— Ductile iron pipe is very strong. It's also heavy and relatively expensive. It may be specified in areas where settlement is possible or where it's vital to prevent sewage leaks that can pollute the surroundings — near a municipal well or public water source, for example. A sewer system may use another type of pipe for most of the system, with ductile iron specified only for sensitive portions. Where this is the case, you'll need to add in the cost of transition fittings (special couplers that adapt from one pipe type to another).

Transite— Is also called asbestos cement, abbreviated AC. It's slightly lighter than concrete and has a reputation for brittleness in the small-diameter sizes. But it's still an adequate pipe material when laid and bedded properly. Transite also uses the slip-type gasket.

PVC and resin compounds— These are light and strong in small-diameter sizes. They're sometimes more expensive but easier to install. Lower installation cost may help offset the higher purchase price.

PVC and resin compound pipe usually comes in 20-foot lengths. This is too long for deep trench work. It's harder to dig and prepare a 20-foot digging set. Solve this problem by purchasing the pipe in 12½-foot lengths. These are especially convenient if you're using a 16- or 20-foot trench box.

PVC and resin compound pipe is easy to cut and bevel. This allows precise positioning of tees and wyes, which can speed up production if the service pipes are laid prior to the main line. You'll use slip gaskets with this type of pipe. Be sure your cutting and beveling are accurate or your gaskets will be out of place and your installation will be faulty.

We've discussed the different pipe materials available. Now let's look at how the pipes are joined.

Fittings

Use sewer pipe fittings to change pipe size, pipe type or direction of flow. Common sewer pipe fittings are the tee, wye, ell, plug, and cap.

Tee— A T-shaped fitting that connects pipes of unequal sizes, or changes the direction of the pipe run. Use tees to connect the main sewer line to the service lines. As the main sewer line will always equal or exceed the diameter of service lines, a 6-inch by 4-inch tee would be used for a 6-inch main line with a 4-inch branch line. The smaller size is always the size of the branch line.

Wye— A Y-shaped fitting that allows the service pipe to enter the main line at a 45-degree angle. You can set wyes vertically and use them with a 45-degree ell. This provides an upright riser commonly used for the cleanout at the end of a line. This allows easy access for tools when it's necessary to clean the line.

Ell— The common term for elbows or bends. The two sizes commonly used for sewer pipe are the 45-degree ell and the 22½-degree ell. They're also sometimes referred to as a fraction of a circle — since one fourth of a circle is 90 degrees. The 45-degree ell is called a 1/8 and the 22½-degree ell is called a 1/16.

These ells come in spigot/bell and bell/bell configurations. The spigot is the plain end of one pipe that goes into the enlarged end of the next pipe to form a joint between the two lengths of pipe. The enlarged end is the bell. The spigot is also known as the "male" end. The bell is the "female" end.

Plug and cap— To avoid water infiltration, all blank and unused connections are plugged or capped. You can use fittings either of the same material as the pipe, or mechanical plugs.

Remember that you'll be testing the system once it's complete. So be sure your plugs and caps are all firmly in place. In fact, it's good practice to place a 2 x 4 service marker against a cap or plug so it can help hold the cap in place during air testing. See Figure 8-4.

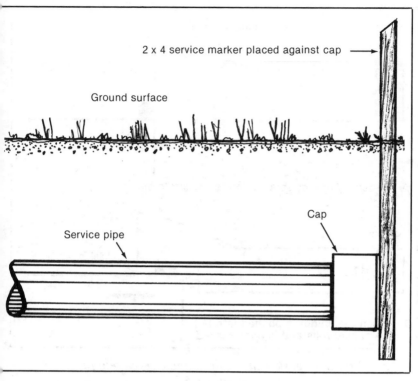

2 x 4 service marker placed against cap
Ground surface
Cap
Service pipe

2 x 4 service marker placed against cap
Figure 8-4

Cleanout— A cleanout assembly consists of a wye, a 45-degree ell and a capped (or plugged) length of pipe rising up to the surface of the ground. In the event of a blockage in the line, the cleanout allows easy access to the line. Cleanouts are usually used where the service line changes direction on private property.

Gaskets

Gaskets provide the seal between individual sections of pipe. All types of pipe require gaskets, except pipe fitted with glue-joints. Rubber compounds are the most common gasket material. Under compression, the rubber is displaced, filling any voids or irregularities between the surfaces of the bell and spigot.

Different types of pipe require different types of gaskets. Be sure you use the correct gasket for the pipe you're installing.

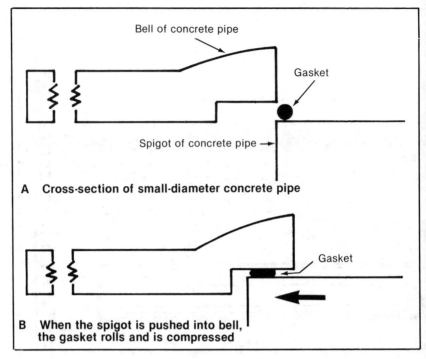

Roll-over gasket in small-diameter concrete pipe
Figure 8-5

Every seal in the line depends on it. For example, gaskets used for pressure pipe rely on internal pressure to force the gasket against the spigot and bell. These gaskets don't work well at low pressures.

There are four main types of gaskets: roll-over gaskets, fixed O-rings, compression gaskets and slip-type gaskets.

Roll-over gasket— Is commonly used for concrete sewer pipe up to 10 inches in diameter. It's usually D-shaped or tear-shaped and is soft. Rolling the gasket into position compresses it.

For a roll-over gasket to work properly, the bell and spigot must be perfectly aligned, and free from dirt, grease or anything that will cause the gasket to slip or not roll over like it's supposed to.

Figure 8-5 shows how to install the roll-over gasket. Set the gasket on the outermost edge of the spigot. Holding the pipe firmly, use your hand and knee to sharply push the spigot into

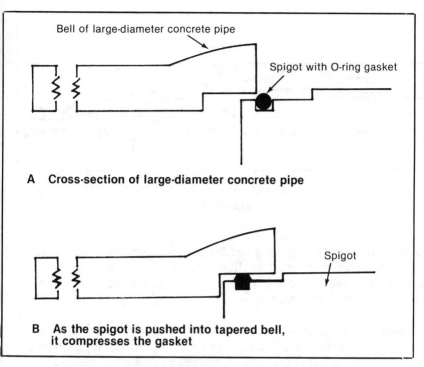

O-ring gasket in large-diameter concrete pipe
Figure 8-6

the bell. Always check the gasket by feel. If the gasket is in the wrong place, you'll be able to feel that it's not seated smoothly inside the bell.

Pay attention to the quality of the finish on the bell and spigot. The gasket will take care of tiny irregularities. But chips or bubble spots in the concrete may result in a bad seal. If a pipe has defects in the finish, either send it back or repair it with grout. Faulty pipe will result in a faulty seal and a joint that leaks. If you're paying for first rate materials, don't accept less.

O-ring gasket— Is used for large-diameter concrete pipe. A groove or indentation in the spigot holds this gasket in place. See Figure 8-6. Use a lubricant to help slide the gasket into the bell. Commonly called "pipe slick" or "soap," pipe lubricant is specially formulated so that it won't harm the rubber com-

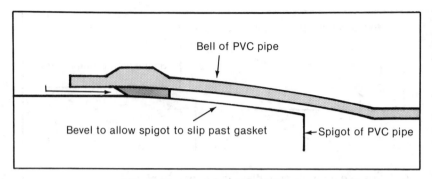

Slip gasket used with PVC pipe
Figure 8-7

pound. In a pinch, you can use household vegetable shortening, which has similar properties. But don't use equipment lubricating grease in place of pipe lubricant. Grease will attack the gasket and eventually erode it.

Compression gasket— Is mechanically compressed and is commonly used in flanges, mechanical joints, water service fittings and pressure sewer lines. The compression gasket is shown in Chapter 7.

Slip-type gasket— Figure 8-7 shows a slip gasket used with PVC pipe. Be sure you insert loose gaskets in the right direction. Some gaskets are marked for easy insertion. Others are hard to figure out. So read the manufacturer's directions carefully. The directions will also tell you how to care for the gaskets. Many rubber gaskets deteriorate in sunlight or exposure to the elements, and need special handling.

Gaskets are used to join the bell and spigot of pipes of the same material. When you join pipes of different material, you'll need a special fitting. Couplers made by Fernco and Calder will accomplish this. They are flexible rubber bands that slip over the spigots of the two pipes and are held in place by adjustable clamps. They're commonly used to join an existing sewer service to the new service at the property line. Where the old service line is 4-inch concrete or clay, it will have an outside diameter (O.D.) of about 6 inches, while the new PVC pipe will have an O.D. of about 4 inches.

Bedding

As described earlier in this chapter, most specifications require that you use bedding material to provide firm, uniform support under the pipe. This increases the pipe's resistance to breakage and deformation. Engineers usually require a minimum of 6 inches of bedding under the pipe. And the bedding has to come at least halfway up the sides of the pipe as well. In many cases the engineer may specify full-cover bedding.

Now that we know the key materials required for sewer system work, let's discuss the equipment you'll need to complete your project quickly and efficiently.

Selecting Equipment

Highly productive excavation is expensive. On a large project, your combined crew and equipment costs may run $5,000 per day. So it's important to select the right piece of equipment for each phase of your project.

Equipment you'll use in sewer system work includes: tracked excavators, wheeled backhoes, backfill equipment, compaction tools, hand tools, laser and optical levels, pumps, and dump trucks. We'll discuss each of these pieces of equipment briefly in this section. For a detailed look at backhoe operation, refer back to Chapter 5. Methods for estimating equipment production rates are given in Chapter 2.

Tracked excavator— There are rare occasions when you'll use a large wheel or chain trencher as your primary excavation tool. But your first choice on most jobs will be the large tracked excavator. It's more versatile and can maneuver around obstructions and cross-lines.

Here's how to select the correct size of excavator. For best production, the trench depth should be at 60 to 70 percent of the maximum reach of the equipment. A machine that can dig down 25 feet will work efficiently in trench depths down to 16 or 17 feet. If the trench goes much deeper than that, the machine will be less efficient.

When an excavator is the primary excavating tool, the work pace of the excavator will dictate the work pace for the other equipment on the job site. If the scheduled production rate is 50 linear feet per hour and it yields 120 cubic yards of soil, the backfill and compaction equipment will have to keep up with that rate.

Tracked backhoe— Most tracked backhoes can be fitted with dipper sticks of varying lengths. In some cases it makes sense to fit a suitable dipper to an existing machine rather than use a different machine. A short dipper stick gives faster cycle times in relatively shallow excavation work. The long dipper stick is better suited to deep trenches. Keep in mind that equipment that's too large will hamper your work just as much as equipment that's too small.

Backfill equipment— Select backfill equipment with slightly more capacity than the job is expected to need. This will ensure that backfilling keeps pace with the excavator. There will be some unavoidable delays in excavation. And there will be times when the excavation goes better than expected. When the excavation rate increases unexpectedly, it's nice to have backfill equipment that can meet the faster pace.

Compaction tools— These tools are described in Chapter 6, Soil Compaction.

Hand tools— They include shovels, a bar for levering the pipe in place, a screwdriver for assembling pipe couplers, a target for the laser, a ladder for access into and out of the trench, a pipe saw for cutting and beveling pipe, and plumbing tools such as a pipe wrench for repair of damaged water services.

Pipe layers tend to be possessive about their shovels, usually preferring a particular size or style. They often put a reference mark on their shovels to help grade out the pipe bedding with a laser. Look at Figure 8-8.

Laser and optical levels— Lasers have revolutionized pipe laying. Use a transit or laser-aligning instrument to set the laser in position and align the beam. Your laser should give the pipe layer a continuous reference point that's precisely on line and grade.

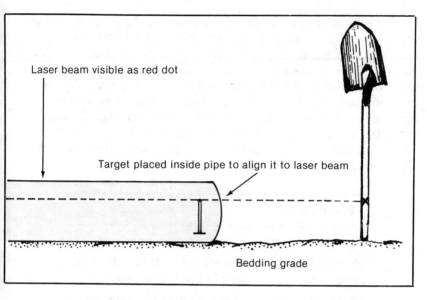

Laser used to align pipe and shovel to grade bedding
Figure 8-8

The laser beam is used to align pipe on both the horizontal and vertical planes. It can also be used to grade out the pipe bedding. The shovel handle is marked as a reference point for the pipe layer to use in leveling out the bedding material before placing the next pipe. Figure 8-8 shows how.

Pumps— If you're working in groundwater conditions, selecting the right pump is important. Capacity is a key consideration. If you can't control the groundwater, you'll never finish the job. In some cases you'll even need 24-hour pumping. When I select a pump, I figure the capacity required and then get a pump that can handle more than that much water. Specs usually confirm that groundwater is expected. But they never say how much. Selecting the right pump can mean the difference between adequate water control and a flooded-out project.

Work trucks— A small one-ton flatbed truck with tool boxes is a good choice because it has more capacity than the 1/2-ton pickup.

Your equipment needs can also depend on the location of the work. If you're working in an open area, the minimum equipment required for continuous sewer pipe laying will be: backhoe, backfill machine (loader or dozer), compaction tools, hand tools, laser and optical levels, and a work truck. If your job is located on a busy street, your equipment requirements may double. You can't interrupt traffic flow through the area, and you'll probably need to backfill the same day you excavate.

Here's a sample equipment list for a pipe laying project in a developed area:

Equipment	Task
Tracked backhoe	Primary excavation tool
Wheeled loader number 1	Material layout, cuts asphalt, supplies pipe bedding
Wheeled loader number 2	Trenches backfill, removes asphalt
Wheeled backhoe	Installs manholes, assists in repairing utilities
Sheepsfoot roller or hoe packer	Compaction
Trucks	As needed

Forming a Productive Crew

Teamwork is the key to good production in sewer system work. The best materials and the best equipment can't make a profit for you if your crew doesn't work together as a team. Even the best superintendent can't finish the job on schedule with a lazy crew. And the best crew will spin its wheels in frustration without a good superintendent. Every member of the team is equally important.

Selecting skilled, efficient workers is essential. But there's more to teamwork than just hiring competent, reliable individuals. Here are three rules for turning a crew of skilled workers into a highly productive, well-coordinated team.

1)   *Provide easy access to materials, tools and equipment.* "If you need it, go and find it" isn't the most productive work method. Make sure your workers have easy access to *all* of the materials, tools and equipment they need. This includes materials that might be required to repair the minor damage that happens occasionally. It's wise to stockpile materials that can be used for repairs. This may seem like an unnecessary expense, but the stockpile will easily pay for itself by preventing costly delays in production.

When you lay out the materials for your crew, include a set of specifications and blueprints that show the layout of the job, including all cross-lines. The more your crew knows about the project, the easier it is for them to perform efficiently. A well-informed crew makes fewer errors.

2)   *Overlap responsibilities.*   Each crew member should have a primary responsibility and several secondary responsibilities. For example, the supervisor, equipment operator and pipe layer should all know the location of the cross-lines. Since more than one person has this responsibility, there's less chance for error. This makes it easier for production to continue without the interruption of damaged cross-lines.

Here's something to look out for when you're assigning responsibilities. Make sure each member of the crew clearly understands the specific tasks he's responsible for. This seems obvious and simple enough — but it's a hitch that causes problems even with experienced crews. A little bit of confusion goes a long way, and it can bring production to a halt in a hurry. Be clear with your crew when you're assigning responsibilities.

3)   *Make safety easy.*   Pipe laying is dangerous work. The safety of the men working in the trenches depends on the skill and judgment of your equipment operators and the above ground crew. Common earth weighs over 100 pounds per cubic foot. So even a small chunk of earth or rock falling only a few feet can cause an injury.

Insist on safety every step of the way. Set the tone by using only safe work practices, proper equipment, material handling, and shoring methods. Insist on hard hats and heavy work boots. It seems like a minor thing. But they've prevented many injuries and saved more than a few lives.

When your crew and equipment cost is running $500.00 an hour, a 15-minute delay will cost you $125.00. Even a little mistake can cost you thousands of dollars. An efficient crew doesn't waste time or make mistakes.

Like your equipment requirements, crew requirements will vary with site conditions. In a new subdivision, the minimum crew for continuous pipe laying will include: backhoe operator, backfill equipment operator, pipe layer, topman (pipe layer's helper), laborer/manhole finisher, and a foreman. On a busy street, your crew requirements can double.

Step-by-Step Installation

Assume you've been awarded a contract and arranged for your materials, equipment and crew. You've already cleared the site. Now what? Let's go step-by-step through a sample sewer system job. We'll begin with the excavation work, selecting a shoring system, then digging and bedding the trenches. Then we'll cut, lay, assemble and align the pipe. We'll install the manholes, test the completed system, and finally, compact the spoil and backfill the trenches.

We're installing a 10-inch sewer pipe in moist loam soil in a residential area. The average depth of the line is 12 feet. There's 6 inches of bedding underneath the pipe and 4 inches of gravel cover on top of the pipe. There are cross-lines at each intersection (at every 150 feet along the main line). Left and right service connections are required every 150 feet. Compaction will be native fill compacted to 95% of standard Proctor. On top of that, we'll put a 1-foot subbase and 2 inches of asphalt.

The first step will be deciding on the shores to use.

Selecting a Shoring System
Contractors who use unsafe or illegal work methods are risking their employees' lives and their own financial health. The best protection against injury is safe, sensible work practices. Only about 1% of industrial accidents can be blamed on unavoidable circumstances or "acts of God." The cause of virtually every accident can be traced to human failure. It may be the fault of management, the worker, or both.

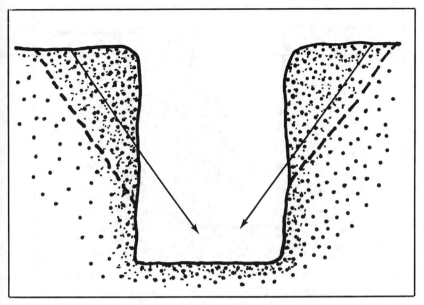

Sandy soil collapses straight down
Figure 8-9

Different soil types, depending on the condition of the soil at the time of excavation, behave differently. Sandy soil will tend to collapse straight down as shown in Figure 8-9. Wet clays and loams tend to slab off the side of the lower trench as shown in Figure 8-10. Firm sand, wet clay and loams should be tight-sheeted using plywood sheets each side of the shoring jacks or with the plywood actually attached to the jacks. Reinforce the jacks with stout 3 x 12 x 8-foot timbers if necessary.

Firm, fairly dry clay tends to crack some distance from the trench wall as shown in Figure 8-11. You can hold dry firm clays and loams using jacks without tight sheeting.

Wet sands and gravels tend to slide into the excavation at about a 45-degree angle as shown in Figure 8-12. They're best dealt with by using a trench box because they're not stable enough to allow time to place shoring jacks or sheets.

Whatever soil type you're working in, it's your responsibility to use a sound shoring system to prevent cave-ins in your trenches. Let's look in detail at the three common methods of shoring: the open ditch, shoring jacks and the trench box.

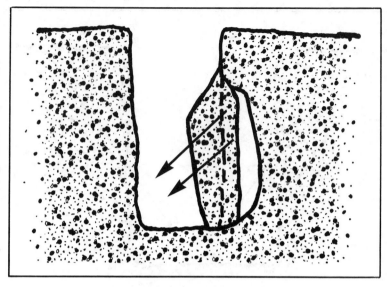

Wet clays and loams "slab off"
Figure 8-10

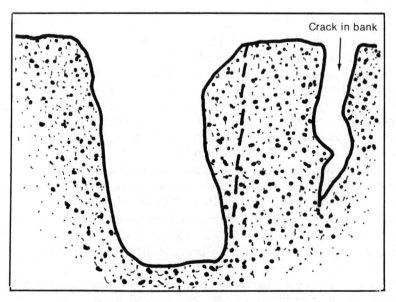

Crack in bank

Firm dry clays and loams crack
Figure 8-11

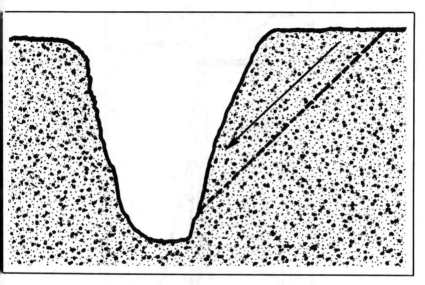

Wet sands and gravels slide
Figure 8-12

Open ditch— In shallow excavations where large-diameter pipe requires a wide cut, you may be able to use the open ditch method. In this method, the walls of the trench are sloped back to prevent the tops of the walls from falling in. Use an open ditch in undeveloped areas and *only in reasonably stable soil*.

Don't use this method where there's a risk of wall collapse. Loose soil and falling rock are the main dangers to a pipe crew. Slabs of material can slide out of the ditch wall. This is more likely when the soil is wet. Water acts as a lubricant and promotes *slabbing*.

Shoring jacks— If the soil is firm clay or loam, shoring jacks are the first choice. See Figure 8-13. They're a quick, efficient shoring system because the excavator can work continuously. Be sure you have enough jacks on hand so you can keep pace with the excavator. You may even need to use more than one worker to move and set the jacks.

Where the ground is slightly unstable, but stable enough so the trench wall will stand while the jacks are being placed, use plywood sheeting between the jacks and the trench wall. This will increase your labor costs but is much cheaper than restoring

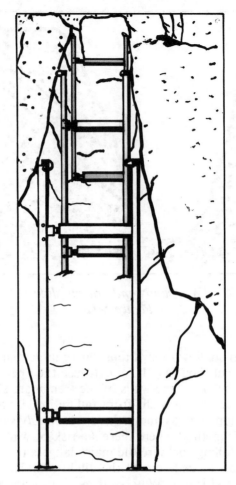

Shoring jacks
Figure 8-13

a cave-in or sending a worker off to the hospital in an ambulance.

You can safely shore a narrow trench in firm ground with shoring jacks. This means you won't need to import a lot of trench backfill material. In damp clays and loams, imported backfill is often specified because the clay and loam are hard to compact. Sand, gravel and crushed rock are the most common imported fill materials. And they're usually expensive because they have to be hauled in.

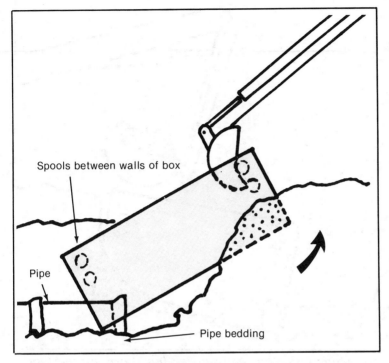

Pulling trench box forward to next section of trench
Figure 8-14

Trench box— For deep trenches and unstable ground, the trench box is the best shoring system. It's a large mobile box with enough strength to withstand the side pressure of deep excavations. Here's how to use it.

Sometimes the trench can be dug ahead of the box and the box simply pulled forward into the next section of trench. But when the ground is extremely unstable, first excavate the top few feet and place the trench box in position to begin. The hoe then excavates *between* the walls of the box. As the backhoe excavates material from between the walls of the box, it shoves the box down into position.

After the pipe is laid in the first section of trench, the hoe pulls the box forward to the next section of trench to be excavated. See Figure 8-14. Then the hoe again excavates between the walls of the box and shoves the box down into position as trench material is removed. Look at Figure 8-15.

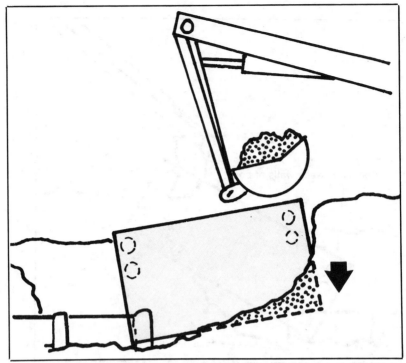

Excavating inside box and shoving box down into position
Figure 8-15

Repeat the procedure until the entire trench is dug.

A disadvantage of the trench box is that there isn't much room for the pipe layer to work, especially when the pipe layer and the hoe bucket are in the box at the same time. Figure 8-16 shows the hoe bucket working in the trench box. The obvious danger is that a worker can be trapped and crushed by the hoe bucket. For this reason, *the operator must never maneuver the hoe bucket unless he is 100% sure of the position of workers in the box.*

As you pull the trench box ahead in unstable ground, the trench will collapse behind the box. A pipe layer can be trapped by loose material as it flows into the rear of the box. The pipe layer should have enough confidence in the box so that he doesn't run out of the box into an unshored trench. It's definitely more hazardous outside the box than inside it, even if the box is temporarily tipped and slammed by falling material.

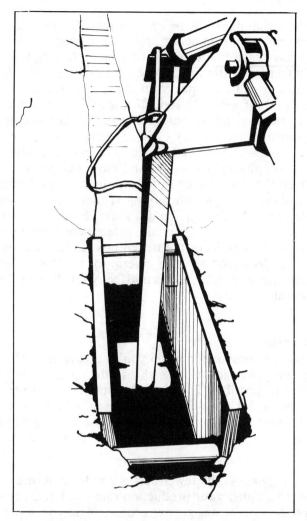

Hoe excavating inside trench box
Figure 8-16

Here's something to watch out for when you're using the trench box. If the box is allowed to sit too low on each side of the pipe you're installing, the side pressure of the pipe bedding can pull the pipe out of position when the trench box is pulled forward. The grademan has to watch the position of the box as it's being installed and instruct the operator to keep the box high enough to avoid this.

There's a variation on the trench box that I've recently heard about. I haven't seen it in action, but it reportedly works very well. It's a steel cage fabricated of I beams, 8 feet high by 8 feet wide by 20 feet long. It's just an open framework with no top, bottom, sides or ends. It's placed in the top of trench excavations and 8-foot by 16-foot steel sheets are driven down each side of the cage as excavation progresses. The strong cage holds the shoring sheets in place. Since the cage and sheets can be handled separately, this system seems to have the advantages of a trench box plus lighter weight and more flexibility.

We've seen that the type of soil determines which shoring system you should use. Which shoring system is best suited to our sample project? The soil is moist loam. Because an open ditch would involve a lot of asphalt replacement, that's out. That leaves shoring jacks or the trench box. Before we make a final selection, let's look at some criteria for successful digging and see how the shoring system can affect your digging production rate.

Digging Trenches

To get maximum production when you're excavating, it's important that your excavator be able to dig continuously with as few delays as possible. The operator must be able to excavate accurately and avoid cross-lines. To choose the best shoring system, you must calculate the trench volume and digging rate correctly. Let's take a closer look at each of these points.

Continuous digging— The fewer delays you have during excavation, the higher your production rate will be. Of course, efficient production means greater profits. Help your operator avoid production delays. Before trench excavation begins, prepare the job site properly and make sure all necessary materials are on hand, including an adequate shoring system.

Staying on grade— A skilled tracked hoe operator can excavate to a tolerance of 0.1 foot (1/10 foot). That takes precise control. But knowing how to maneuver the equipment isn't enough. The operator must be able to clearly see the work, even when there are obstructions. The operator often relies on the hand signals of the pipe layer to guide him around obstacles.

The pipe layer uses the laser beam as a reference point. When the excavator bucket is close to the desired grade, the laser beam will appear on the bucket. If there are high spots, the beam will appear on the unexcavated dirt. Using the beam as a guide, the pipe layer can direct the operator to a very fine tolerance. The pipe layer's guidance should prevent excavation errors and boost production.

Tracked hoe operators have a difficult job. It's hard to excavate to grade accurately when the hoe bucket is 15 to 30 feet away from you. A grademan is essential for the tracked hoe operator. Different operators and grademen use different hand signals to communicate with each other. The exact signals aren't important — just so they're communicating.

Avoiding cross-lines— The grademan has to give hand signals to avoid cross-lines because the operator often won't be able to see them. A backhoe can break most cross-lines with ease. The pipe crew should help the operator avoid damaging cross-lines by marking their location well ahead of the excavation equipment.

If the operator knows in advance the exact depth of the cross-lines, he can safely excavate above and below that depth. The pipe layer can help by signaling the depth and exact location of the lines when the hoe bucket gets close. The operator should slow the machine so it can be stopped in time if the cross-lines aren't exactly where they're supposed to be.

Teamwork is essential. An alert grademan keeps track of cross-lines so his operator can concentrate on getting maximum production. If either member of the team isn't doing his share, both will fail. Stress teamwork with your crew.

Calculating trench volume— The size of the pipe and the type of shoring system you use will determine the volume of material you take out of the trench. Even when the pipe diameter is the same, a change in the shoring system makes the trench a different size.

First let's find the volume for a vertical wall trench. Look at Figure 8-17. Here's the formula. Note that all dimensions are given in feet.

$$\frac{\text{Width x depth x length}}{27} = \text{B. CY}$$

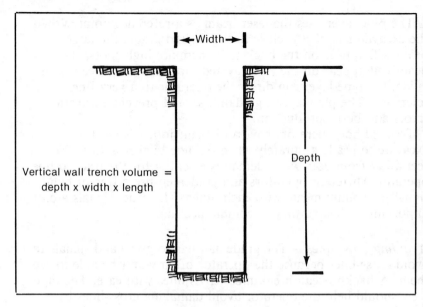

Volume of vertical wall trench
Figure 8-17

Next, we'll find the volume of a trench with sloping sides. There are two ways to do this. You can in effect "square off" the trench by mentally drawing a line straight up from one side of the bottom cut. You're cutting off one triangle and adding one of equal size to the other side. Look at the dark and the light triangles in Figure 8-18. Consider that the width is only the distance from one edge of the trench to the imaginary line. Multiply that by the depth and length, just as in the formula above.

Or you can average the width by adding the top width and the bottom width, and dividing the answer by 2. Look at Figure 8-19. Here's the formula:

$$\frac{\text{Top width} + \text{bottom width}}{2} = \text{average width}$$

$$\frac{\text{Average width} \times \text{depth} \times \text{length}}{27} = \text{B. CY}$$

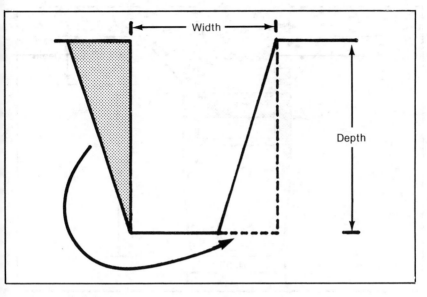

Volume of a trench with sloping sides, method 1
Figure 8-18

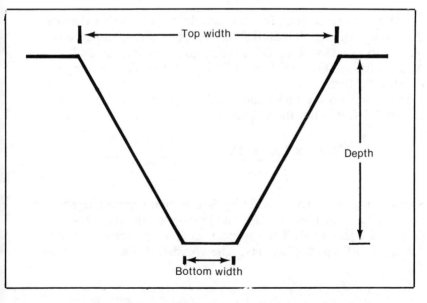

Volume of a trench with sloping sides, method 2
Figure 8-19

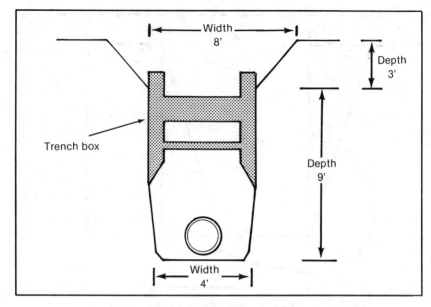

Volume of a trench with trench box
Figure 8-20

Using the dimensions for our sample project, let's calculate the volume of trench material for the open ditch, trench box and shoring jacks. Remember we're going to dig, lay and backfill for 3,000 linear feet of 10-inch sewer pipe. Average depth is 12 feet.

If we use shoring jacks and lay the pipe in a vertical wall trench 2.5 feet wide, the volume is:

$$\frac{3000' \times 2.5' \times 12'}{27} \quad = \quad 3{,}333 \text{ B. CY}$$

Suppose we use a 4-foot wide by 6-foot high trench box for our project. Look at Figure 8-20. The bottom of the trench is 9 feet deep and 4 feet wide. The top portion is 3 feet deep, and the "squared off" width is 8 feet. The volume of the trench would be:

$$\frac{(8' \times 3') \ + \ (9' \times 4') \ \times \ 3000'}{27} = 6{,}667 \text{ B. CY}$$

If we used an open trench and sloped the walls back at a 1/2 to 1 ratio, the trench volume would be:

$$\frac{9 \ \text{x} \ 12 \ \text{x} \ 3000}{27} \ = \ 12,000 \ \text{B. CY}$$

Notice the dramatic increase in bank cubic yards of volume: 3,333 B. CY for shoring jacks, 6,667 B. CY for a trench box, and 12,000 B. CY for an open trench with sloped sides. And the open trench would entail greater surface restoration. You don't have to be a genius to realize it will be easier and cheaper to move 3,333 than 12,000 B. CY. But the choice between shoring jacks and the trench box is less obvious when you consider you'll have to employ men to constantly move the shoring jacks. If you use two extra men to move the jacks, the hoe could have virtually uninterrupted trenching.

But there's one more calculation we have to consider before we decide. Let's look at how trench volume affects the digging rate.

Calculating digging rate— Now that we know the trench volume for each of the three shoring systems, we can calculate the actual digging rate. We'll use a tracked backhoe for this job. Chapter 2 tells you how to calculate production rates for that machine. For purposes of this example, we'll assume a cycle time of 30 seconds, based on the difficulty of maneuvering around the obstacles on this job. That's 120 cycles an hour.

Next we need to apply the swell factor to convert the B. CY of trench volume to the L. CY of bucket capacity. Back in Chapter 2, Figure 2-4 gives a swell factor of 40% for moist loam. That means our 3,333 B. CY would swell to 4,666 L. CY. The 6,667 B. CY excavated for the trench box would swell to 9,334 L. CY.

The hoe is fitted with a 2 L. CY bucket. Since we're digging in moist loam, the struck capacity will be about 2.2 L. CY. If we multiply 120 cycles an hour by 2.2 L. CY, we could dig 264 L. CY per hour — *before* we adjust for efficiency and operator ability. Let's use the standard 0.83 factor for efficiency, and 0.70 for operator ability. That reduces the production to 153 L. CY per hour.

Now the calculation looks like this:

For the shoring jacks:

$$4,666 \div 153 = 30.5 \text{ hours}$$

For the trench box:

$$9,334 \div 153 = 61 \text{ hours}$$

It may seem needlessly complex to analyze production rates this way, but how else are you going to make production decisions that maximize your efficiency and your profits? Now you have the data you need to decide whether to use shoring jacks or a trench box for this job.

Bedding the Trenches

Bedding provides firm support for the pipe. This helps protect the pipe from breakage or becoming deformed. The bedding technique you use will depend on whether you're working in an open area or a developed area.

In open areas— There are three common methods of placing bedding in trenches in open areas:

• Front loader: You can use a front loader to dump the bedding material over the edge of the trench.

• Bedding box: A small backhoe is all that's needed to tend a bedding box. That should free up your front loader for other work. Or trucks can dump directly into the bedding box. The hoe shoves the box, repositioning it along the trench as needed. That may eliminate the need for a wheeled loader. Just pace your deliveries of bedding material to keep up with the other equipment.

• Rock holes: The third method of placing bedding in trenches in undeveloped areas is to excavate rock holes. These are shallow pits dug at intervals along the line. The pits are filled with bedding material. There's likely to be some waste with this method, but the waste may cost you less than providing the equipment necessary for the other methods. Also look at how many tasks you've given to your primary excavator. It may not be cost-efficient for him to interrupt continuous excavation to handle the bedding material.

If it's practical to use rock holes, calculate the cubic yards of bedding material needed per linear foot. Space the rock holes at intervals that will allow full truck loads to be dumped into the holes. If minimal bedding is needed and that spacing would be too far for the excavator to travel to pick up the rock, use half the distance.

The excavator can revolve 180 degrees, track toward the rock holes, load, and transport the bedding rock into place without the use of a loader. At today's cost, a wheeled loader and operator will cost upward of $50 an hour. While using rock holes causes some waste of bedding material, it can be cost effective because it saves the cost of the loader supplying bedding. You can waste a lot of rock for $400 a day.

Cutting the Pipe

Accurate measuring and cutting are essential in sewer system work. The product of careless measurements and cutting will be leaks. And leaky sewer pipes won't pass the air test. Here's how to cut small- and large-diameter sewer pipe accurately.

Small-diameter pipe— Figure 8-21 shows how to use a disc saw to make a square cut in small-diameter pipe. Don't make the common mistake of rolling the pipe to mark it all the way around before cutting. Here's how to make an accurate cut.

Begin by moving the saw into cutting position at a 90-degree angle to the pipe, as shown in section A. Holding the saw in this position, make a deep cut into the pipe (section B). *Now* you can roll the pipe forward a little bit, as shown in section C. Using your first cut as a guide, insert the saw again and make the second cut. Section D shows how. For the final cut, roll the pipe forward again and repeat the procedure.

Large-diameter pipe— For large pipe, you'll need to mark the pipe all the way around before cutting. But don't just mark one spot and hold a pencil on the spot and roll the pipe to complete the marking. Make several marks on the pipe by *measuring from the end of the pipe.* Measure each mark separately to make sure they're all equidistant from the end of the pipe. Position your saw at a 90-degree angle to the marks and begin cutting.

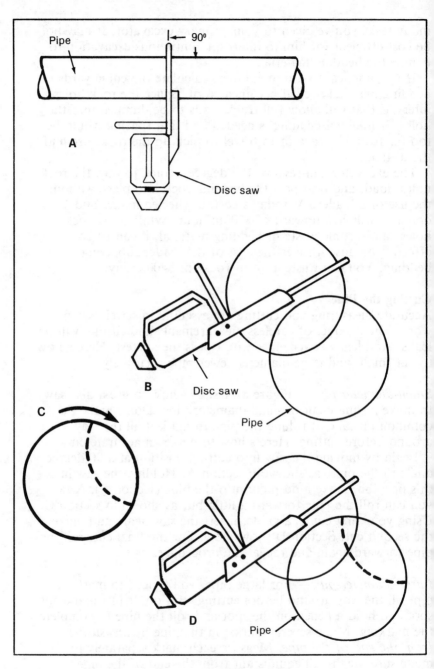

Cutting pipe with disc saw
Figure 8-21

The Pipe Layer
According to many wage scales, pipe-laying wages are little
more than common labor wages. But the job of pipe laying
involves much more skill, experience and responsibility than that
of a common laborer.

The job involves working with expensive equipment: lasers,
surveying levels, and transites. It requires physical labor, of
course, but more often it requires directing the heavy equipment
used in deep sewer installation. It also requires an understanding
of mathematics, elementary surveying and plan reading.

A deep sewer installation project involves the use of tons of
heavy equipment costing hundreds of thousands of dollars to
move thousands of cubic yards of earth to install pipe to a
tolerance of one or two hundredths of a foot. The pipe layer
has primary responsibility for making it all go right.

But that's not all. The pipe layer does all of this in a
dangerous environment. Even with the best safety practices, it
only takes one small mistake or misjudgment to cause an injury
or death. One of the basic rules of industrial safety is to stay
out of areas where overhead objects can fall on you. Few
workers would deliberately stand under the eaves of a house
with framers or roofers working above them. But by the very
nature of their work, pipe layers routinely work in trenches 5 to
25 feet deep with other activities going on above them. Besides
this obvious risk, the trench itself is a potential death trap. This
make safe shoring absolutely *vital*.

In order to get the pipe in the ground safely and profitably,
the pipe layers, the topman and equipment operators must work
together as a well-coordinated team. There's no room for slips.
The work *must* happen as planned. Without the required
coordination, it'll be dangerous, it'll be expensive — it'll be a
fiasco!

Depending on the excavation rate, the pipe layer may or may
not remain in the trench while the hoe digs the next set. If he
does, he's at risk from error on the part of the hoe operator.
An overfull hoe bucket incorrectly maneuvered can drop
hundreds of pounds of earth or rock on him. He's also
threatened by error on the part of workers on the surface of the
ground. It's this risk that makes the topman a vital part of the
team.

The topman's role— The topman carries and hands down tools,
slings the pipe if it's being machine handled, checks pipe grade,

fuels and maintains pumps, cuts and prepares pipe and fittings, and generally assists both the pipe layer and hoe operator.

Remember, the counterweight of the large tracked hoe swings out and away from the machine's centerline in the opposite direction of the boom and dipper. Because this happens out of sight of the operator, he'll often depend on the topman to let him know whether or not the counterweight is clearing power poles, traffic signs, mail boxes or other vertical obstructions. So it's the topman who acts as another pair of eyes for the tracked hoe operator.

The topman has a similar duty when pipe bedding is dumped over the edge of the trench. The loader operator probably won't be able to see the pipe layer. The bedding must be dumped accurately or it will displace small diameter pipe aligned by the pipe layer. So it's the topman who transfers the pipe layer's directions to the loader operator. While the pipe layer, operator and topman each have different jobs, each job is vital.

Laying the Pipe

You can lay pipe either by hand or by machine. Whether you're laying by machine or by hand, always handle pipe with care. Most pipe is strong, but any pipe can be damaged by carelessness.

Hand laying— Most small-diameter pipe is laid by hand. Pipe weighing up to about 180 pounds per length can be laid by hand, if necessary.

Concrete pipe is usually "thrown" into the ditch by the topman. This technique only works when the walls of the trench are nearly vertical. The topman picks up the pipe and swings it (in an upright position) out over the trench. Then he releases it to fall, bell first, onto the pipe bedding. Here's an important rule for all pipe throwers: *Don't throw the pipe until the pipe layer asks for it.* The pipe layer will be preparing the pipe bed and may accidentally walk under a falling pipe.

Never stand pipe up vertically near the edge of a trench. It can fall over into the trench and onto the pipe layer. Long sections of PVC pipe (12 to 20 feet) should be handed down into the trench. And remember that PVC pipe can be fragile at low temperatures and requires extra care.

Machine laying— Sometimes the tracked excavator digging the trench is used to lay the pipe. But this cuts into production time. Two other machines often used for laying pipe are the wheeled backhoe and the crawler pipe layer. A skilled backhoe operator can maneuver his machine very efficiently on a dry, firm spoil pile. And the machine's capacity will allow it to lift and set pipes weighing 1,500 pounds or more. The crawler pipe layer may cost more per hour, but it'll get the job done quicker. Here's how to calculate pipe laying production rates.

Let's say we're using a wheeled backhoe to lay 21-inch by 6-foot concrete sewer pipe 10 feet deep into a 2½-foot wide trench. Assume it takes 12 minutes to dig a two-pipe (12-foot) set. Remember that a set is a section of trench that can be excavated with the hoe in one position. The time required for a wheeled hoe to dig, bed and lay the set might look like this:

Dig set	12 min.
Bed set	2 min.
Hook, lower and lay pipes (2 minutes per pipe)	4 min.
	18 min.

This makes your hourly dig, bed and lay rate:

$$\frac{60 \text{ min. per hour}}{\text{min. req. to dig, bed, lay set}} = \text{sets per hr x length of set} = \text{number ft. of trench per hr.}$$

$$\frac{60 \text{ min.}}{18 \text{ min.}} = 3.33 \text{ sets} \times 12 \text{ ft.} = 39.96 \text{ (or 40) ft. per hr.}$$

If the hoe only has to dig and bed (no laying), the rate will be:

Dig set	12 min.
Bed set	2 min.
	14 min.

If pipe is laid by a machine other than the hoe, the hoe production rate will look like this:

$$\frac{60 \text{ min.}}{14 \text{ min.}} = 4.29 \text{ sets} \times 12 \text{ ft.} = 51.48 \text{ (or 51) ft. per hr.}$$

The hoe will be producing at a rate of 51 feet per hour. This is 11 feet per hour faster than if it had to lay the pipe in addition to digging and bedding it. The savings from the

Assembling pipe with steel bar
Figure 8-22

additional 11 feet per hour may be enough to pay for a separate pipe laying machine. Check the cost of each work method before you commit yourself to it.

Assembling the Pipe
The assembly technique you choose will depend on the type and size of pipe being laid. Small-diameter concrete pipe uses roll-over gaskets. You can shove the pipe together by hand. Small-diameter pipe made of non-PVC material uses slip-type gaskets. Use a steel bar to shove this type of pipe into position. The steel bar gives you additional leverage. If the pipe is in the right place, you can join the two sections of pipe in one smooth motion. See Figure 8-22.

Medium-diameter pipe (10 to 21 inches) requires a different assembly technique. You can suspend it from a laying machine and swing it into position (stab joining). This works well with transite and ductile iron pipe. Or you can use a backhoe to lay it in position and shove it carefully into place.

Joining the main line and the service lines— The pipe material will determine whether you install the service lines or the main line first. If the pipe is concrete or clay, the tees are an integral part of the pipe. Location of the tee can vary only about a foot. The service pipe must be installed after the main line. But if you're using PVC pipe, the placement of the tees is more flexible. This has several advantages:

• The service pipe can be installed before the main line.

• The wheeled backhoe installing the service pipe can install it more quickly and easily.

• If the service pipe is installed ahead of the main line, the service pipe trench is dug from the property line back to the main. Service pipe can be laid to grade and in a position that allows easy hookup as the main line passes through.

Here's how to calculate the position of the lower end of the lines you install first. Where the tee is set 45 degrees off vertical, the vertical separation will equal the horizontal separation. A 45-degree service ell will be needed. Where the tee is set 22½ degrees off vertical, the horizontal separation will be 0.4 of the vertical separation. A 22½-degree ell and a 45-degree ell will be needed for service ells. See Figure 8-23.

Having an offset between the service pipe and the main line works fine as long as there's a vertical separation of at least 2 feet between the two lines. When the vertical separation is less than 2 feet, it's hard to get enough "flex" to make a good hookup. In this case, you may want to use the engineer-approved *no-hub coupler* or a *bell-by-bell slip coupler* to join the service pipe.

When you're ready to join the service pipe and the main line, the line you're hooking up to should already be well filed and beveled. If it's only been factory-beveled, bevel it some more. This will make hookup easier. You can use a gasoline or electric abrasive disc (or special beveling tools) to speed beveling on

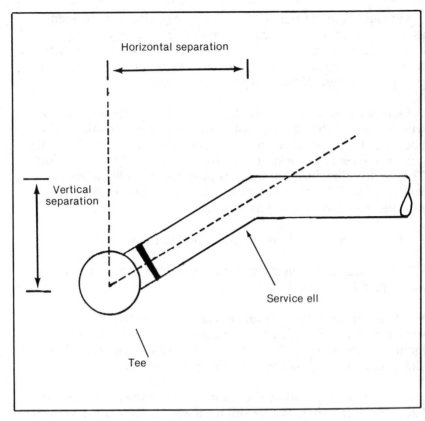

Finding the position of the service pipe ell
Figure 8-23

large-diameter pipe. For small-diameter PVC service pipe, a good rasp will do the job. It's always easier and safer to do your beveling above ground and before the pipe is installed.

We've talked about the advantages of installing the service lines first, but there's a disadvantage, too. If the service pipe is installed first, the main line crew has to hook up each service as it comes to it. This can slow production of the expensive main line crew and equipment. It may be more efficient for the main line crew to just install the riser pipe of the service. That's the section of service pipe that rises from the depth of the main pipe up to a level where you can install the service pipe to the property line. The service line is usually from 5 to 8 feet deep at the property line.

A smaller, cheaper service crew consisting of a wheeled backhoe, operator and pipe layer can follow behind, extending the services to the property line. It's common practice to dig backwards from the property to the riser pipe. The main line crew marks the end of the riser pipe with a 2 x 4.

The operator of the wheeled backhoe has a difficult job: First, the service crosses under curbs and gutters, sidewalks and the other utility main lines. Second, he's digging directly toward the fragile riser pipe. If he damages the riser pipe or pulls it out of the main, it's a disaster.

There's a safe way to dig toward the riser. Look at Figure 8-24. The letters show the sequence of digging. Take section A out first, then carefully excavate sections B and C. The section of dirt marked D is moved last. It's shoved down and away from the capped riser. The service pipe layer must closely watch the digging to help avoid damage. In most cases the wheeled hoe operator can't see into the trench. Hand digging completes the exposure of the capped riser and it's ready for hook up.

As always, teamwork is vital. A good service team can install service pipes at the rate of nearly one an hour. There are some tricks that help production. Rather than excavating completely under sidewalks and cross lines, it may be possible to push a pipe or steel bar underneath the obstruction. Even a 2-inch diameter hole is enough to get the expander through. An expander can be made out of a length of steel cable attached to a 5 or 6-inch diameter plug. It's pulled through the smaller hole to expand it enough to get the 4-inch service pipe through. Make plenty of 4-inch couplers available to the service crew so that pipes can be quickly and easily joined.

Aligning the Pipe

Once the pipe is laid and assembled, it must be aligned and brought precisely to grade. A laser beam is the most common method of aligning pipe. But you can also use an optical level. Here's how to align the pipe and correct any alignment errors that might occur.

Laser alignment— A transit or laser-aligning instrument will set a laser in position and help you adjust the beam to line and grade. The laser projects this beam for several hundred feet. The pipe layer uses the beam as a reference point. Wherever the beam of light is intercepted, it shows up as a red dot. The red

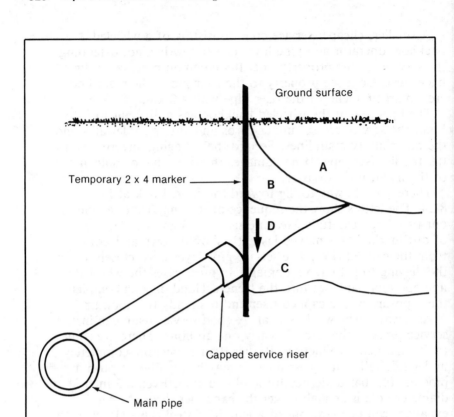

Digging toward a riser pipe
Figure 8-24

dot of the laser beam is the reference point that's used to align the pipe on both horizontal and vertical alignment.

But the laser isn't perfect. If you've ever used a high-power telescope, you know that differences in air temperature can distort the line of sight. Differences in air temperatures in a pipeline can deflect the laser beam.

Changing temperatures cause the most trouble when you're working in extremely hot or extremely cold weather. Pipe that's been in the ground for several days will have a moderate temperature because it's insulated against the heat and cold. But lengths of pipe that are strung out along a proposed pipeline may be much warmer or cooler.

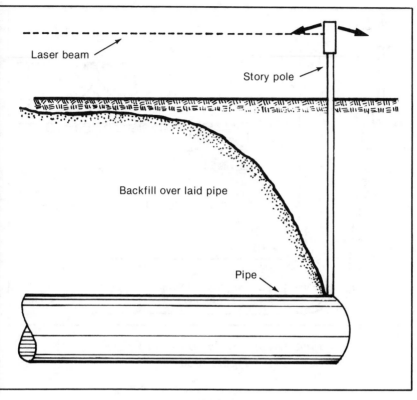

Transferring grade with a story pole
Figure 8-25

Most laser manufacturers offer an air blower that can be set up adjacent to the laser. The blower moves air through the pipeline. This helps maintain a more even temperature. But the beam may still wander off its true line and grade. Before laying the next joint of pipe, it's wise to recheck the position of the beam in the pipe you just laid. If the beam is in a different position, it means the beam is wandering. If the beam is moving, monitor the pipe grade and alignment very closely. Check the grade every 20 feet and the alignment every 50 feet.

Laser problems— Here's one solution to a problem laser beam: Set up the laser above ground and use a story pole to transfer grade to the pipe. Look at Figure 8-25. Let's say you establish that the laser beam is exactly 10 feet above the top of the pipe.

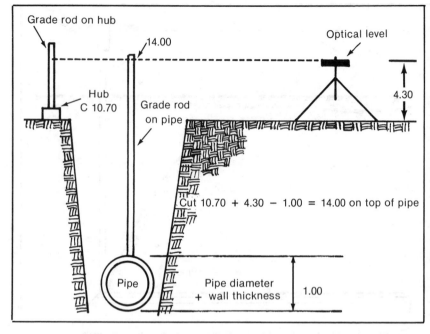

Grade rod on hub

14.00

Optical level

Hub
C 10.70

Grade rod
on pipe

4.30

Cut 10.70 + 4.30 − 1.00 = 14.00 on top of pipe

Pipe

Pipe diameter
+ wall thickness

1.00

Finding rod reading on top of pipe
Figure 8-26

Dial the correct grade into the laser. The rest of the pipe must also be 10 feet below the laser beam. This works well in relatively shallow trenches. In deeper trenches, there's the danger of a false reading if the story pole isn't held perfectly vertical.

Personally, I don't feel comfortable with a laser beam that gives a hint of trouble, or with the story pole method. It's much better to correct the problem beam by using blowers, or to "shoot in" each pipe with an optical level.

I covered laser set-up and grade checking in Chapter 3, but let's recap. The engineers provide hubs and a cut figure at intervals of 25, 50 or 100 feet along the pipeline and at each manhole. Set up an optical level, establish its height above the hub, then add it to the cut figure to find the *required rod reading at flow line*. Look at Figure 8-26. If you set the grade rod on top of the pipe, you have to deduct the diameter of the pipe plus the thickness of the pipe wall. This gives you the *required rod reading on the top of the pipe*.

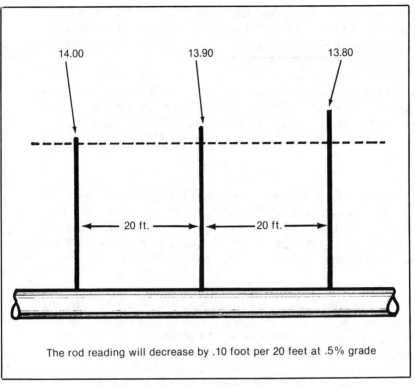

14.00   13.90   13.80

|← 20 ft. →|← 20 ft. →|

The rod reading will decrease by .10 foot per 20 feet at .5% grade

Checking grade with an optical level
Figure 8-27

Using this information, you can calculate the *required rod reading on the top of subsequent pipes.* If you're laying pipe at 0.5% grade, you're raising the pipeline 50 feet per 10,000 feet, 5 feet per 1,000 feet, 0.5 foot per 100 feet, or 0.05 foot per 10 feet. See Figure 8-27.

You can just leave the optical level in the same spot and elevation, deducting 0.05 foot for each 10 linear feet of pipe. When the distance between the optical level and the rod becomes too great, simply move the level ahead, establish the rod reading from the last pipe, and make subsequent adjustments from that rod reading.

To check grade quickly and easily, talk to the pipe crew and translate the grade into grade difference per joint. For example, at a 0.40% grade, the pipeline rises 0.1 foot per 25 feet. That translates to 0.05 foot per 12.5-foot joint of pipe.

Checking the horizontal alignment is especially important in the first section of pipe. If the pipe is out of alignment, it'll show up near the beginning. And it's easier to correct alignment at the beginning than at the end.

The laser's horizontal alignment is especially critical when you're installing small-diameter (6-inch to 8-inch) pipe. When the pipe is that small, there are only a few inches for correction before the beam is interrupted by the wall of the pipe. Laser alignment is less critical when you're installing large-diameter pipe.

Correcting alignment errors— If you find your pipe is out of horizontal alignment, three common ways of handling it are: laying to a curve, installing an additional manhole, or starting over. Get full approval by the engineer's inspector before you make a decision on how to correct the problem.

• Laying to a curve: Some inspectors will allow a slight curve in the pipe. But laying pipe to a curve is difficult because the laser can't be directed around the curve. This means you're constantly resetting the laser or "shooting in" with an optical level as each pipe is laid.

• Installing an additional manhole: You can change the direction of the line by installing an additional manhole. If it's the contractor's error that caused the problem, it's not likely he'll get paid for the manhole. But it's not always the contractor's error. The engineer's surveying team may have made an error that caused a misaligned pipe. In that case, you can demand payment for the extra work and for any loss incurred because of it.

• Starting over: You can always begin again. This can be an expensive option. But if you catch the error early enough, it may be less costly to start over than to continue laying to a curve or to install an additional manhole.

Installing Manholes
Manholes are used to change the direction of lines and at intersections of lines coming from different directions. Excavating for manholes requires special techniques. Teamwork between the operator and pipe layer is especially important.

When a pipeline goes straight though the manhole, excavation is pretty simple. Where the pipeline makes a 90-degree turn, however, the operator should try to do as little excavation as possible to minimize the amount of surface restoration that's needed. The equipment operator will have to rely on the pipe layer's directions to excavate the manhole accurately.

In high groundwater, manhole excavation can be slow and tedious. You'll probably have to excavate a couple of feet below pipe grade to allow enough depth for the manhole base. If high groundwater is a problem, digging that extra foot or two below pipe grade can draw a lot of water, even if water is being controlled at the level of the pipe bed.

Groundwater in sand and gravel makes them almost fluid. An inexperienced operator may make the mistake of trying to excavate too quickly. A big hoe can scoop out the spoil very quickly. But any splashing in the hole will cause more subsidence. Excavation becomes a never-ending process as banks undermined by the water collapse into the hole.

Figure 8-28 shows a workable technique. If you're digging in groundwater, follow these steps:

1) Try to get a pump below the water table to dry out your excavation.

2) Keep the pump intake clear by surrounding the pump intake in drain rock, or by placing it in a slotted pipe.

3) Drain rock helps stabilize running sands and gravels. Use it.

4) If you don't have shoring wide enough to adequately shore a manhole excavation, open it up. Slope the excavation walls back to a safe angle.

Blockages are rare in a well-designed, properly installed sewer system. But access is still needed to be able to get into the system easily to inspect, monitor, clean and repair it. All gravity sewer systems require manholes at specified intervals. The most commonly used manhole material is precast concrete. Figure 8-29 shows a section drawing of a typical manhole. A manhole has three main parts: the section, cone and base.

Section— Precast manhole sections are usually 4 or 6 feet in diameter and 1, 2 or 3 feet high.

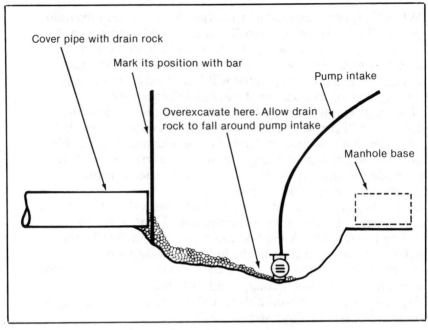

Excavating for a manhole
Figure 8-28

Cone— This is the top section of the manhole. The cone is labeled in Figure 8-29. It's usually 2 or 3 feet high. There are two types of cones, *eccentric* and *concentric*. The eccentric cone has a top opening set off to one side of the circular section. The concentric cone has a top opening in the center of the circular section. Use the concentric cone for the shallower manholes. Flat-top manholes are also available for very shallow trenches and special purposes such as lift stations.

Base— This is the bottom section of the manhole. Precast bases are available with or without precast channels. Precast bases must be set to a precise elevation. In shallow manholes and on firm ground, it's easier and more economical to use ready-mix concrete for the base, and to set the precast sections on top of it. Here are three ways to handle this:

• Set a 1-foot precast section up on blocks and pour the base inside of and around it.

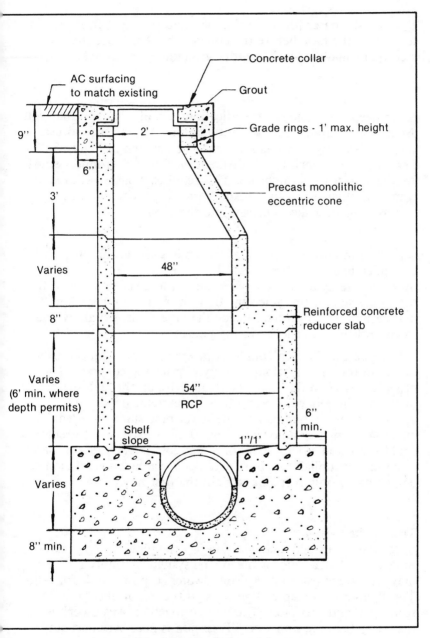

Typical manhole section
Figure 8-29

• Pour the base, form the channels and place a 1-foot precast section on the base before the concrete hardens. Use this method in shallow work and dry ground.

• In deep wet trenches, specifications usually demand that you use a base of drain rock as a foundation. And be sure to use adequate shoring, such as the trench box shown in Figure 8-30. Removing the trench box is difficult in deep wet ground because you have to lift it straight up. This requires a large backhoe or crane. Another way you can shore the deep trench is with interlocking steel sheets driven into the ground.

Even with adequate shoring, controlling the groundwater while the concrete sets is often a problem. The best method is to overexcavate by a couple of feet and set a sewer pipe vertically off to the side of the hole. Surround it with drain rock and place a pump suction hose inside the pipe. This will allow you to control the water level during pouring.

Here's a tip. Place a couple of layers of 6-mil visqueen over the drain rock foundation before you pour the concrete. This provides a seal so that the water can't rise up through the base of the manhole if your pump fails. In extreme conditions, double up on the pumps and the sewer pipe sump — a leaking manhole in high groundwater conditions is expensive to repair. I could show you several manholes that cost the contractor many thousands of dollars to construct. Deep manhole installation can be profitable only if it's done right the first time.

Testing the System
Specificatons usually demand you test the pipelines to make sure there are no leaks into or out of the system. A faulty system may let sewage leak out or groundwater seep in. Look for both. Throughout this chapter I've stressed the importance of teamwork and precision. The two go together. Teamwork is necessary for precision work. Precision is necessary because everything you install will be tested and may have to be guaranteed for at least one year.

No project is complete until it's been tested and inspected.

Courtesy: GME Trench Boxes

Manhole trench box
Figure 8-30

That's a measure of your workmanship. Several different tests are used on sewer lines. Any or all of them may be used on the lines you install. Here are the most common sewer line tests.

Visual testing— A common visual test is *lamping,* shining a powerful flashlight through the line. An inspector looks toward the lamp from the opposite end. The inspector can tell if the line is straight and true and on grade — and will see any curves where the line should be straight. Any puddles caused by low spots in the line will also be obvious.

A T.V. camera can also used to make a visual inspection, especially on curved sections that don't permit direct observation of all pipe between manholes. A camera can examine every inch of the interior of a pipeline. Ponds, puddles, high spots, low spots, joints, infiltrating water, dirt and silt in the pipes will all show up. The inspection can also be recorded: Potent evidence if you didn't do something right.

The mandrel test— Some pipe material may flatten or deflect under backfill and compaction. To make sure that it hasn't, some inspectors demand a mandrel test. This involves pulling a mandrel through the line. The mandrel has a slightly smaller diameter than the pipe and will get stuck if there's been any deflection.

Water testing— There are two common water tests: the infiltration test to find out if water is infiltrating into the pipe, and the exfiltration test to see if water is leaking out.

For an infiltration test, the line is sealed off. Water infiltrating the line will accumulate in downstream manholes. The source of the infiltration can be detected by lamping or T.V. inspection.

For the exfiltration test, portions of the line are isolated and the manholes filled with water. If the water level drops significantly, there's probably a leaky manhole. Figure 8-31 shows how the exfiltration test works. It's easier to place and remove a plug located in a manhole downstream from the manhole being tested.

If this isn't possible, you may have to have an extension hose fitted to the plug so you can drain the flooded manhole you're testing. Look at Figure 8-32. There's an air hose extension to allow the air plug to be released when the manhole is flooded. The inflatable plugs are tied to the manhole steps to prevent the plugs from being washed down the pipe when the water is released.

Concrete pipe will absorb some water during the test. Don't get excited about the water level dropping during the first 24 hours. It always will. But any fall in the water level after that will probably be due to exfiltration rather than absorption.

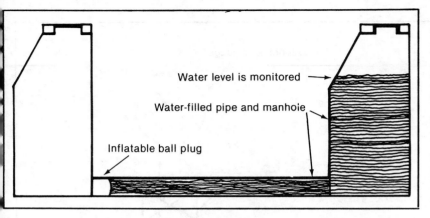

Exfiltration test
Figure 8-31

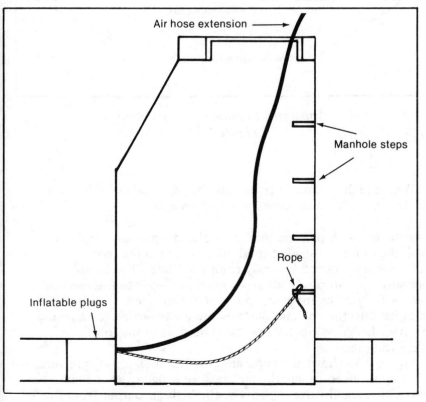

Extension hose fitted to inflatable plug
Figure 8-32

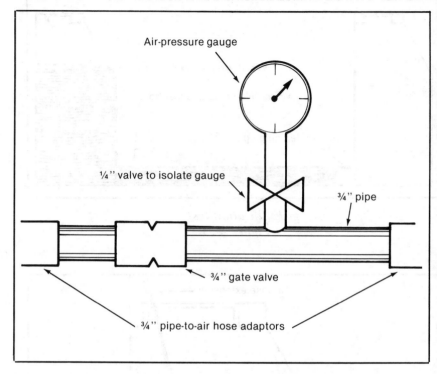

Apparatus to monitor air pressure
Figure 8-33

Where high groundwater outside the pipe makes infiltration likely, an exfiltration test may not be required.

The air test— A pipeline that can hold low-pressure air for a period of time is usually sound. The air test is the most common test applied to newly completed lines. For an air pressure test, all pipe ends are capped and sealed. Each section can be isolated by inflatable plugs at the manholes. Then air is pumped into the line and the pressure is monitored for a period of time. Exact details for the test should be in the job specifications.

Figure 8-33 shows the apparatus used to monitor air pressure. It's important to isolate the air pressure gauge from the main air line with a small valve. Otherwise the high air output from the compressor will damage it.

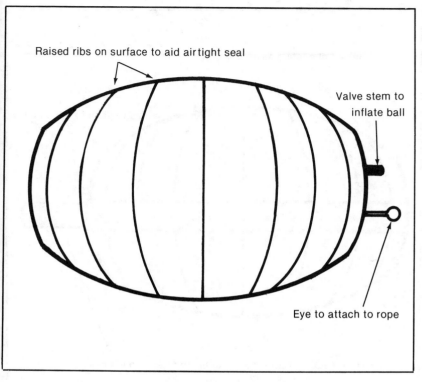

Raised ribs on surface to aid airtight seal

Valve stem to inflate ball

Eye to attach to rope

Simple inflatable plug for testing lines
Figure 8-34

Inflatable plugs are used for both air and water testing. Figure 8-34 shows a simple inflatable ball. Figure 8-35 shows a second type of inflatable ball that has a hollow tube through it. Air is pumped through the tube in the ball to the pipeline. These large test balls are very expensive. You can fabricate larger diameter plugs from wheel centers and inner tubes.

Even though the air test is usually low pressure, from 4 to 10 psi, it can still be dangerous. Always brace large diameter plugs with lumber to keep the air pressure from dislodging them. And use common sense. If you're removing a test ball in a manhole, there isn't much room to get out of the way if it blows from the air pressure. Remember, 10 psi means 10 pounds of pressure *per square inch of surface.* A 24-inch diameter plug has an area of 52 square inches. That's 520 pounds of pressure. Make sure air pressure in the line is back to normal before removing the plugs.

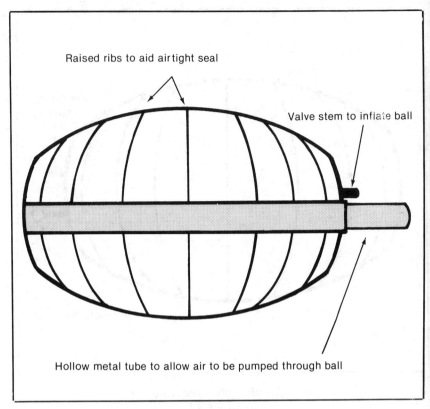

Test plug with hollow tube
Figure 8-35

Cleaning the Lines
Silt and dirt in pipe reduces flow capacity. Cleaning is
sometimes necessary. There are two ways of doing this. First,
you can use a special flushing ball. It's similar to the test ball
but has grooves around the outside which jet water past the
outside of the ball. To use it, rapidly flood a manhole. Rising
water forces the ball and dirt downstream. Then retrieve the ball
by reeling in the rope attached to the ball. Pulling the ball back
in causes water to jet past the grooves, dislodging silt and dirt
accumulated at the bottom of the pipe.

The second method is to hire a cleaning service with
specialized equipment that can flush the lines with high pressure
water.

If you need to get a rope through a line, either float a light nylon string downstream or blow a string through with a special parachute-umbrella specially made for the purpose. If neither of these works, I've seen a $10 battery-powered toy truck do the job. A nylon string was attached to the bumper and the truck was sent down the line. A few minutes later the truck came out at the other end, with line attached. This works especially well in small diameter pipes, providing the lines are reasonably clean, of course.

Backfill, Compaction and Surface Restoration

Backfilling and compaction for your sewer line projects are explained in Chapter 6. Surface restoration can involve landscape restoration, fence removal and paving.

Asphalt surfacing is best left to specialist subcontractors with the equipment and know-how to patch or resurface. The same is true for concrete work. If you have the people and know-how, by all means do it yourself. If not, stick to doing what you do best.

If you have to do much surface restoration, let me recommend an excellent tool that will save hours of hand work with a rake. Fit a 4 to 5-foot wide toothless bucket on a wheeled backhoe. A bucket like that will skim spoil off lawns easily. A substitute is a 4-foot wide by 1-foot high by 1-inch thick steel plate bolted or welded to a standard hoe bucket.

9

ESTIMATING SURFACE EXCAVATION

Pipeline contractors occasionally need to estimate the cost of grading or leveling land. If you have to bid a job that includes some grading, the formulas used for figuring soil quantities back in Chapter 2 won't be enough. For trench excavation, the quantity of soil to move equals trench length times width times depth. Even if you have to average the depth of the trench, the formula remains the same.

The calculations are a little more complex if you have to compute the volume of soil to be moved when leveling land for a building or a road. The design engineers will usually show only the finish elevations and the existing elevations. It's up to you to find the quantity of material to be moved. The shape to be excavated may be irregular: curves, hills, gullies and depressions. Multiplying the width times the depth times the length won't work. That's why this chapter is important. It will explain the formulas that do work when bidding a grading or leveling job.

Figuring quantities for grading and leveling is more complex. But the work itself may be simpler. On a pipeline job you can have a narrow right of way, rock, groundwater, and difficult compaction problems. Grading and leveling will be done in relatively open areas — and only the top few feet of soil may be involved.

To estimate general excavation, you must be familiar with volumes, grids, shapes and averages.

Remember that a cubic yard of soil is the volume that would fit in a perfect cube 3 feet on each side (27 cubic feet). Of course, a cubic yard of soil could also measure 9 feet by 3 feet by 1 foot. Or it could cover 108 square feet (12 square yards) at

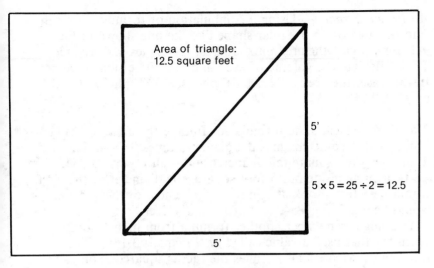

Area of a triangle
Figure 9-1

a depth of 3 inches — or any other regular or irregular shape you can imagine. It's measuring the irregular shapes that's difficult when calculating volumes.

Before we discuss how to do this, let's take a look at the formulas used to calculate areas and volumes. *Area* measures a flat surface: It's two dimensional. To compute volume, we must first find the area, then multiply by the depth.

The square— A square is a simple geometric shape. To calculate its area, we simply multiply its two dimensions by each other: for a 5-foot square, 5 feet times 5 feet equals 25 square feet. When you multiply a dimension by itself, it's called *squaring* the number. In math symbols, it looks like this: 5^2.

The rectangle— You find the area of a rectangle the same way: base times height equals area.

The triangle— If we divide a square as shown in Figure 9-1, we have two triangles, each having an area of half the original square: that's the base times the height divided by 2. If the original square had an area of 25 square feet, then each triangle has an area of 12.5 feet.

An irregular shape— Using a combination of formulas, we can find the area of an irregular shape like the one shown in Figure 9-2. First divide the area into squares, rectangles and triangles. Then calculate the volume of each and add these sums together. In this case, the area is 72,600 SF plus 4,750 SF plus 7,700 SF, or 85,050 SF.

The circle— Here's the formula for finding the area of a circle: Multiply the diameter times the diameter times 0.7854. To state it another way, square the diameter and multiply by 0.7854. If we fit a circle inside our 5-foot square, the diameter of the circle would be 5 feet. Look at Figure 9-3. The area of the circle would be 5 x 5 x 0.7854, or 19.635 square feet.

You might remember another formula from high school geometry that used a value called *pi*. Pi equals 3.1415. The formula is: Area equals *pi* times the radius squared. You can also use *pi* to find the circumference of a circle: diameter times *pi* equals circumference.

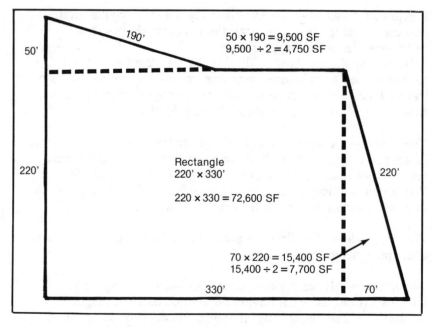

Finding the area of an irregular shape
Figure 9-2

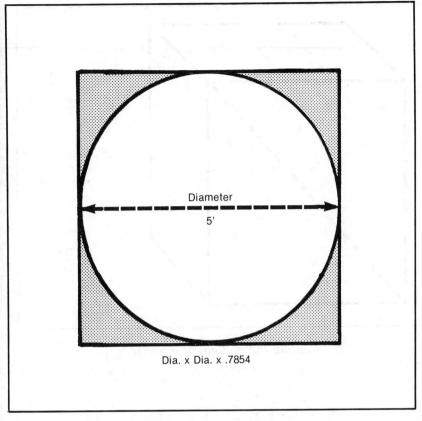

Dia. x Dia. x .7854

Finding the area of a circle
Figure 9-3

Use *pi* if you like, but I prefer my formula. It's easy to find the diameter of the circle, and easy to remember 0.7854. Notice in the illustration that the circle fills a little over 3/4 of the square.

The cube— A cube is the three-dimensional form of the square. To calculate a cubic volume, you just multiply the height by the width by the depth. If the 5-foot square we've been talking about becomes a 5-foot cube, its cubic volume is 125 cubic feet (5 x 5 x 5 = 125). In math symbols, that's 5^3.

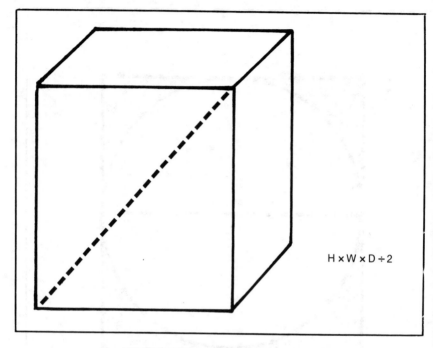

H x W x D ÷ 2

Finding the volume of a wedge
Figure 9-4

The wedge— If we slice a cube into two wedges (Figure 9-4), they will each have a volume of half the cube, or height times width times depth divided by 2.

The cylinder— If a circle has depth, it becomes a cylinder. It's easy to find the volume if you remember the formula for a circle. Just add its depth to that formula: diameter times diameter times depth times 0.7854.

The cone— It's not obvious at a glance, but it's a fact that a pyramid inside a cube, as shown Figure 9-5, will have a volume of precisely one third of the volume of the cube. Yes, I know pyramid building contracts have declined in recent years. But consider this. A cone (Figure 9-6) is identical in volume to a pyramid. Now do you begin to see the applications? You can find the volume of a pile of sand fairly easily. The formula is: Area of base times height divided by 3.

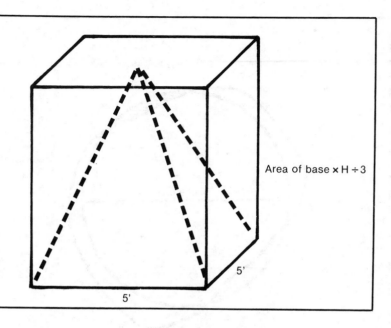

Area of base × H ÷ 3

Finding the volume of a pyramid
Figure 9-5

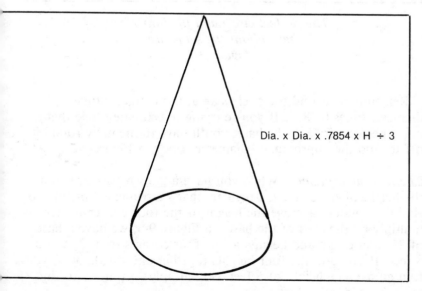

Dia. x Dia. x .7854 x H ÷ 3

Finding the volume of a cone
Figure 9-6

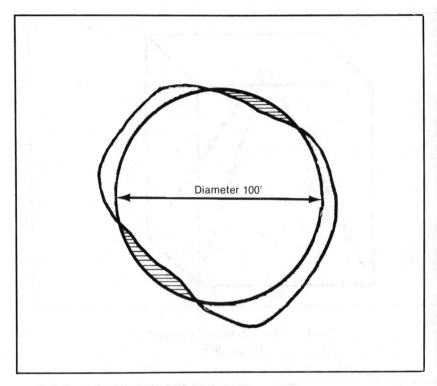

The perfect circle and the imperfect
have about the same area
Figure 9-7

Remember, to find the circle's area, use diameter times diameter times 0.7854. If you're trying to estimate a pile that's not a perfect circle at the base, you'll have to mentally round it off to find the approximate diameter. Look at Figure 9-7.

The truncated prism— When you have a known base area but the height of the shape is irregular, that's a truncated prism. To find the volume, average the height of the four corners. Then multiply by the area of the base. In Figure 9-8, we have a base of 25 square feet and heights at the four corners of 4, 3, 5 and 8 feet. If you add the four heights together and divide by 4, you find an average height of 5 feet. So multiply 25 times 5 to find the area, 125 cubic feet.

And that leads us to the subject of averaging.

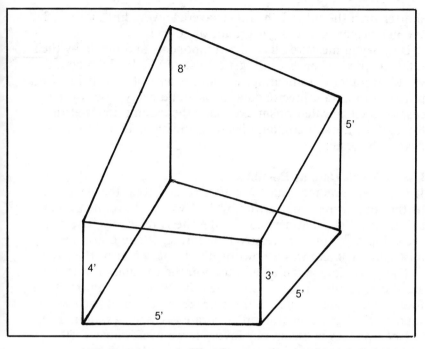

Finding the volume of a truncated prism
Figure 9-8

Averaging in Estimating Earthwork

An average is the sum of the separate components divided by the number of components measured. If we have a man who's 7 feet tall and a shorter man who's 5 feet tall, their average height is 6 feet (7 plus 5 divided by 2 equals 6.)

It doesn't matter if an average is calculated from a total of three or 300 components. The formula remains the same. If there are 300 people, find their combined height and divide by 300. It's obvious that a mistake in the addition leading to the combined height of the people or a miscount of the number of people you're averaging will give a false average.

It's also important to remember that if we were to take the combined height of only 100 out of the 300 people, their average height would probably give an indication of the average height of the 300 people. This only works if the 100 people are representative of the larger group. If they were proportionally shorter than the others, then their average wouldn't indicate the average height of the larger group, of course.

Unless you measure all of the components and divide by their total, there's a chance you'll get a false result. In the practical world of quantity estimating, we have to accept this. It just isn't practical to obtain precise data in each and every case. In the case of the truncated prism, we made the assumption that the top surface doesn't contain hills or depressions that would distort the result.

Use as Much Data as Possible
If we take a grading plan for a 10-acre site, total the elevations of the four corners and divide by 4 — will we have the average elevation? Well, that depends. If the site is relatively flat, with even slopes, it could be relatively accurate. It could also be very misleading if there was a bowl or hill in the center of the site.

The more data gathered and used in the calculation, the more accurate the answer. If we sectioned the 10-acre site into 100-foot squares and calculated the average elevation of each square, we'd get a much more accurate average elevation. The law of averages works in our favor. If we overestimated 50 percent of the squares, the chances are the other 50 percent would be underestimated.

That kind of averaging is often valuable in estimating earthwork quantities. You can find an average volume quantity by multiplying the area by the *average* depth or cut.

What else can we do by averaging? Let's say you need to calculate the volume of water retained in a dam. The chances are that the valley in which the dam is built is extremely irregular in shape. It's a nightmare of a mathematical problem. Well, maybe not. If you have the resources to measure the width and the depth at regular intervals, you could easily calculate the average area at each 100-foot, 500-foot, or 1,000-foot interval along the dam water. Taking the measurements at 100-foot intervals will give you a more accurate answer.

Look at Figure 9-9. We've taken the measurements and averaged them to find the depth and width. Now all we'd have to do to find the volume is multiply those by the length. Or we could calculate the volume contained in each 100 feet of length and add the results together, again arriving at a reasonably accurate conclusion.

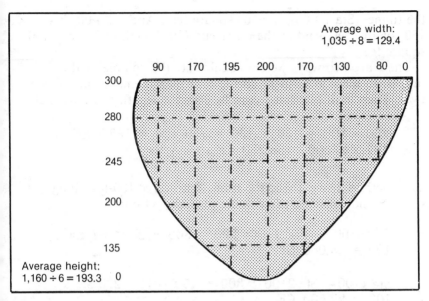

Finding average height and average width
Figure 9-9

The Average End Method

What does finding the amount of water in a lake have to do with estimating earth moving? Not much — except the method. Earth movers call it the *average end method*. And they're probably finding the average height of a hill rather than the average depth of a lake. In mountainous country where the highway cuts through hills and over filled-in gullies, the designers calculated cut and fill quantities by averaging depths at regular intervals.

Because people make mistakes in long strings of calculations, it's usually more accurate to divide up the earth quantities into smaller sections. Calculate the volume for each section, then add the results together to arrive at a total volume.

Let's look at another example of the average end method. Assume you're looking down a tunnel. It's 300 feet long, but the height and width vary along its length. You're going to measure the height and width, or area, of the tunnel at the beginning, the end, and two places in the tunnel. At point 1 (Sta 0 + 00), the area is 200 square feet. At point 2, 100 feet into the tunnel (Sta 1 + 00), it's 300 square feet. At Sta 2 + 00, it's 150 square feet. And at the other end (Sta 3 + 00), its area is 100 square feet.

There are two ways we could calculate the volume of this tunnel. In method 1, just add the four measurements and divide by 4. Then multiply that figure by 300, the length of the tunnel:

> 200 + 300 + 150 + 100 = 750 ÷ 4 = 187.5 SF
> 187.5 x 300 = 56,250 CF

In method 2, we divide the tunnel into three 100-foot long sections and calculate the volume of each section:

> Sta 0+00 — Sta 1+00: 200 + 300 ÷ 2 = 167 SF x 100 = 25,000 CF

> Sta 1+00 — Sta 2+00 : 300 + 150 ÷ 2 = 225 SF x 100 = 22,500 CF

> Sta 2+00 — Sta 3+00 : 150 + 100 ÷ 2 = 125 SF x 100 = 12,500 CF

> 25,000 + 22,500 + 12,500 = 60,000 CF

Which is the correct answer, 56,250 cubic feet or 60,000 cubic feet? The truth is, we just don't know. We're not using sufficient data to determine the volume precisely. If we took measurements at each 25 or 50 feet, we'd get a more accurate answer.

The average end method is reasonably accurate if you arrive at it by using adequate data. But it's good practice to cross-check it by using the formulas outlined earlier in this chapter. Wherever a shape resembles a wedge, a square, a rectangle or cylinder, use the appropriate formula.

Balancing Cut and Fill

Excavation for most buildings requires moving soil over a large area, but at relatively shallow depths. The engineer will usually try to *balance* the quantities of cut and fill so no soil has to be moved from or to the site. This isn't always possible, of course. For example, if all the earth on a site were evenly spread out, it might raise the site too high above the surrounding ground. Or the level could turn out to be too low for access roads, sewer lines and storm drains.

Calculate these quantities carefully. There's plenty of opportunity for serious error. And mistakes can be very expensive. On a 10-acre site, a 0.5-foot error in grade level amounts to an 8,000 cubic yard mistake. If this quantity of material has to be bought and trucked to the site, the cost of that error will be several thousand dollars.

Estimating volumes of irregular shapes is hard enough. But harder still will be forecasting the percentage of swell and shrinkage as soil is excavated and compacted. We covered that subject in Chapter 2. Figure 2-4 is a chart of swell factors for different kinds of soil.

When estimating general excavation, you'll compute the volume in bank measure (usually abbreviated *B. CY*), convert this figure to loose volume (to find the quantity that has to be moved) and then convert back to compacted volume. It's little wonder that few estimators feel comfortable working with earth quantities.

There are dozens of types of soils. They can be wet, saturated, dry, or any condition in between. Most of them can be compacted to varying degrees of maximum density. Each soil type has subdivisions that are difficult to account for. A loose cubic yard of sand from one source may weigh 2,500 pounds dry. Another, finer, sand may weigh 2,600 pounds dry.

Damp earth of any type is usually easy to handle and process. Mud is virtually useless until it dries. Very dry fine-grained dirt is difficult to work with unless water is added. Again, you're forced to make judgment calls.

The final steps in your estimate will be converting volumes to manhours and equipment hours and then multiplying by the appropriate hourly costs.

The Grading Plan

Look at Figure 9-10. It shows a typical grading plan for a residential subdivision. You'll see two terms on most grading plans: *existing grade* (abbreviated E.G. or sometimes O.G. for original grade); and *finish grade* (F.G. or sometimes F.S. for finish surface).

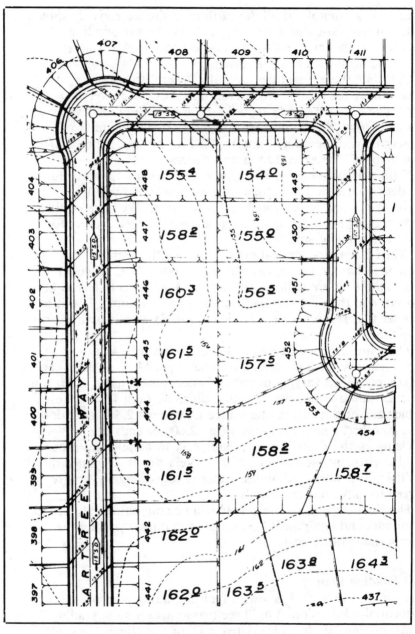

Typical grading plan
Figure 9-10

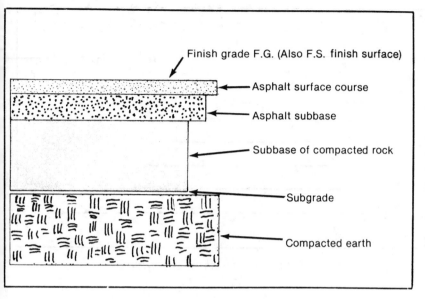

Section view of a road
Figure 9-11

Finish grade doesn't always mean the level of the soil when excavation is complete. For example, a parking lot may have the same E.G. and F.G. But that doesn't necessarily mean you won't do any excavation. The F.G. probably refers to the finished grade of the *parking lot surface* when the job is complete. That level is well above the top of the soil. You would have to excavate for the base material and the thickness of asphalt. The term *subgrade* identifies the surface elevation prior to placing any surfacing material. Figure 9-11 shows a section view of a road subgrade. The subgrade is the finish grade minus the asphalt surface course, minus the asphalt subbase and minus the subbase of compacted rock.

Reading Contour Lines
Contour lines connect points of the same elevation. See Figures 9-12 and 9-13. The curving broken lines on the grading plan in Figure 9-10 are contour lines. In relatively flat areas, the contour lines are widely spaced. On steep slopes, contour lines will be close together. You'll use these contour lines to calculate the difference between existing grade and the finished grade.

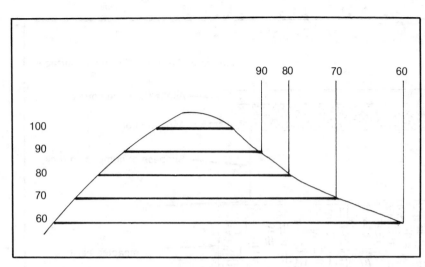

Contour lines show elevation
Figure 9-12

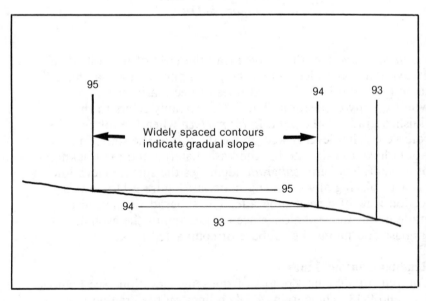

Spacing of contour lines indicates
steepness of slope
Figure 9-13

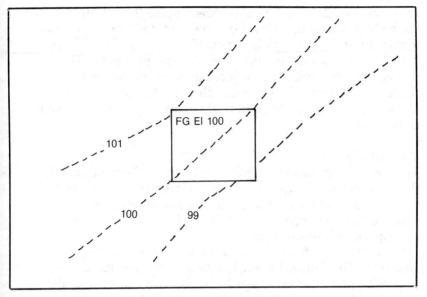

Contour lines and finish grade
Figure 9-14

Where a solid line ties into an existing contour line, it shows the proposed contour. Look at Figure 9-14. In this case, the solid lines indicating FG El 100 extend from contour line 101 to contour line 99. It indicates a cut at the 101-foot level and a fill at the 99-foot level.

There's a saying that a cost estimate is a series of errors carried to the precision of two significant digits. There's some truth in it. You have to make some approximations when calculating earthwork quantities. There's no need to figure quantities down to the last shovel load. But neither do we want to compound errors by using approximations that are inaccurate.

My advice is to use a simple quantity estimating system that may have an error of several percent rather than work with a complex procedure that may be extremely accurate but also error prone. A simple estimating system is less likely to yield a serious mistake because it can be checked quickly. Always cross-check where possible using different methods.

Some rounding will be necessary, of course. For example, 1/3, or 1 divided by 3, is 0.333333. We round it down to 0.33.

In the case of 2/3, we round 0.666666 up to 0.67. Rounding to the nearest hundredth of a foot won't get you into trouble. A hundredth of a foot is only 1/8 inch.

Earthwork estimating also requires some judgment. The procedure is to divide the project into sections and estimate each section separately. The estimator chooses the size of each section. The smaller the section, the more accurate the estimate is likely to be.

The Grid System
To estimate excavation quantities on irregular ground, divide the plan into grids (squares), then compute the volume in each grid. If you make a misjudgment or error in one grid, the error is limited to one portion of the site and not compounded through the whole estimate.

Look at Figure 9-15. Contour lines show the existing elevation. The finished grade is indicated by the figures in the box. We know, however, that between two contour lines the elevation changes up or down. It's a reasonable assumption that a point half-way between the contour lines will have an elevation of half the difference between the contour elevations. By gauging the difference between the lines, we can approximate the existing elevation where the F.G. is shown.

Draw regular grid lines— The process is to draw grid lines at regular intervals, say 100-foot squares. Use the scale marked on the plan to do this. Many plans that you get will have been reduced in size after having been drawn. This makes the indicated scale wrong. Check to find the actual scale by measuring some known distance on the plan. Pencil in your estimated existing grade in each corner of each grid. Designate each grid A1, A2, B1, B2, C1, C2 and so on. Note that this has been done in Figure 9-15. If you're concerned about writing and drawing on the plans, overlay the plan with tracing paper for the grid.

Establish the variables for each grid— You need to establish these variables:

1) The average existing grade in this particular grid

2) The average finish grade in this grid

3) The difference between these two elevations

4) The amount of adjustment you must make to find subgrade

5) The volume of material to be excavated or filled to reach
    the required subgrade

Let's follow this procedure to find the volume of soil to be
moved in Grid D3, Figure 9-15. This grid will become the
northwest corner of a parking lot. The contour lines show a
slope descending from the northeast to the southwest. Contour
line EL 97 nearly intersects the northeast corner of Grid D3.
The northwest and southeast corners are very close to contour
line EL 96. The southwest corner of grid D3 is approximately
half-way between contours 95 and 94.

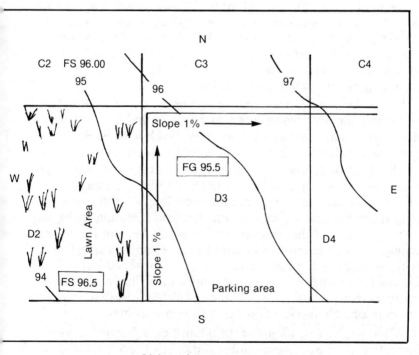

Using the grid system
Figure 9-15

We'll use this information to establish *average existing elevation.* To make checking easier, let's write down the numbers as we come up with them. On a separate sheet or columnar pad, identify the grid as D3, identify each corner and the existing elevation in each corner:

Grid	NE EL	NW EL	SE EL	SW EL	Ave existing EL
D3	97.0	95.7	95.9	94.5	= 95.77

To find the average elevation, add up the four elevations of each grid corner and divide by four. In this case, it's 95.77. We'll round it up to 95.8.

Always look at your average to see if it makes sense. If your average came out 99.0, it should be obvious that there's an error. The highest corner is 95.9. How could the average be greater than this?

Remember that the average of 95.77 was rounded up to 95.8. And notice that we estimated the elevations of grid corners. I estimated the northwest corner at 95.7 by noting the position of the grid corner in relation to the contour lines 95 and 96. It looked like the grid corner was about 7/10th of the distance from contour 95 to 96. So I put down 95.7.

I used the same process on the southwest corner, estimating it to be half the distance between contours 94 and 95. So I wrote down 94.5. The southeast corner must be close to 95.9. The northeast grid corner is nearly on contour line 97.

This kind of estimating will probably alarm a mathematical purist — but we have to be practical. To find the exact elevation at each corner would take too long — and it wouldn't help much because we're assuming the grid elevation is the same as the average of the four corners. It won't be. But if you use enough grids, the errors will tend to cancel out, leaving you with very nearly the correct volume to be moved. Most excavation estimators agree that you're more likely to have a serious error because of a mistake in calculations than in the assumptions about elevation between contour lines.

When you're calculating earth quantities, reliability is more important than extreme accuracy. Since it's probably impossible to predict the precise swell or shrinkage of the earth to within 5%, there's no point in getting excited about absolute precision in your data. Instead, concentrate on recording corner elevations in a form that is easy to check. And worry about the big error that can put you out of business.

Look again at Grid D3. In big letters it tells you that you're dealing with a parking area. A detail someplace on the plans will show what's required to build the parking area. In this case, the detail is Figure 9-11.

Detail C3/sh2/sh20 means you look for Detail C3 referred to on sheet 2 (sh2) on sheet 20 (sh20).

For this exercise, let's take subgrade as F.G. minus 1.0 foot.

The line connecting the box labeled *F.G. 95.5* with the northwest corner gives us the finish grade at that point. The slope arrows and note tell us the parking area slopes from the south to north at 1% and from the west to the east at 1%. This gives us the information we need to establish F.G. in the southwest and northeast corners. A slope of 1% in 100 feet is 1 foot. So we write this down:

Grid	NE EL	NW EL	SE EL	SW EL
D3	94.5	95.5		96.5

The southeast corner of Grid D3 is a mystery. It's probably revealed in data shown in another grid. If not, we can take the three known finish grade elevations and total them to calculate an average finish grade elevation:

$$94.5 + 95.5 + 96.5 = 286$$
$$286 \div 3 = 95.5 \text{ Ave. F.G. EL}$$

From Grid D3, we have established the existing grade and the finish grade. Where the existing grade is greater, we subtract the finish grade from the existing grade. Then, because the area will have 1 foot of material under the finish grade, we subtract this to arrive at the subgrade. (Finish grade minus thickness of imported fill equals subgrade).

In our case, the thickness of imported fill is 1 foot. Subtracting that from the average finish grade, we find that 94.5 is the subgrade elevation. Now subtract that from the average existing grade:

$$95.8 - 94.5 = \text{Cut } 1.30$$

Minimize errors by writing your answer in each grid. Use different color pens for each important entry. You might use black for average existing grade, blue for average finish grade and green for the average subgrade.

Calculate quantities for each grid— We now have the information needed to calculate quantities. The grid area is 100 feet by 100 feet, or 10,000 square feet. Multiply that by the average cut of 1.3 feet:

$$100 \text{ x } 100 \text{ x } 1.3 = 13,000 \text{ B. CF}$$

That gives us the amount of the cut for that grid in bank cubic feet. To convert that into bank cubic yards, divide by 27:

$$\text{B. CF} \div 27 = \text{B. CY}$$
$$13,000 \div 27 = 481 \text{ B. CY}$$

At last we have a cubic volume figure to work with.

When you've followed this same procedure for each grid, just accumulate the sums to find total job volume. Then the all-important questions: *How to do it, how long will it take, and how much will it cost?*

You may have recognized the grid we just estimated as a truncated prism. If the parking lot had a circular planting area, you could use the formula from the beginning of the chapter to figure the area, and the grid method to figure its depth. If there were a triangular lawn area, for example, use the formula for the area of a triangle and multiply by the average depth.

As you calculate the volume in each grid, you'll begin to see a pattern evolve. There will likely be areas of cut and areas of fill. Use color to illustrate this: Color all cut areas one color and all fill areas another color. Now you can clearly see how the earth has to be moved.

When you've finished with all grids, you'll have a total of cut cubic yards and fill cubic yards. The difference between the two is the excess of dirt you'll have to haul away or the amount of fill dirt you'll have to import.

The zero line— The point where the cut areas and the fill areas meet is a line that needs neither cut nor fill. This is sometimes

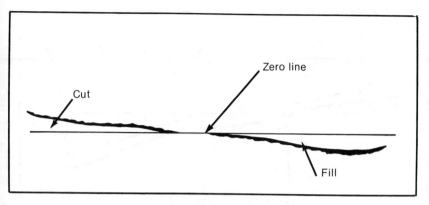

The zero line
Figure 9-16

called the *zero line*. You can see the zero line dividing the cut and fill sections in Figure 9-16.

At the zero line, where the existing grade and the finish grade are the same, there's no allowance for stripping the topsoil or undesirable matter from the site. If you have to strip 0.5-foot of topsoil from the site, the zero line will move toward the cut area. See Figure 9-17.

If no stripping is needed, the zero line will be the balance point between cut and fill. In the case of a simple excavation, as shown in Figure 9-15, the zero line will be relatively straight. If the existing site is very irregular, the zero line will be erratic. In some cases, there may be more than one zero line.

Coloring the grading plan will help you see the pattern of excavation and fill. The distance from the zero line to the edges of the cut and fill areas will help you determine the most suitable equipment to use. For a short haul, you'll probably use dozers. For a longer haul, use self-loading scrapers. Long-distance hauling requires trucks. You can use a grader for very slight cuts and fills. We'll discuss equipment selection more fully in the next chapter.

Minimize the Risks

Nothing is certain in the earth mover's world. It's up to you to minimize and rationalize the risks. Any successful excavating

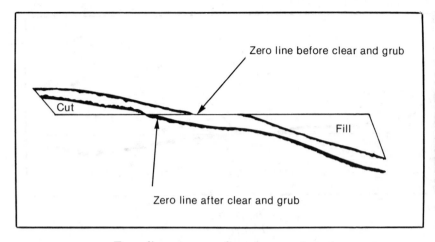

Zero line moves after clear and grub
Figure 9-17

contractor takes risks on a daily basis. But don't be a foolhardy gambler. Try to:

1)   Identify the risk

2)   Predict the likelihood of an unwanted outcome

3)   Place a dollar amount on it

With an earth moving contract, there are many variables that can alter the outcome — and your profit. They include the swell and shrinkage of the materials you're handling, the weather, the condition of the soil, and the amount of clearing and grubbing. Let's look at these points one at a time.

Swell and Shrinkage
On most jobs you'll get a boring log in the bid package. Figure 9-18 shows typical boring log entries. These logs will usually be on a plan sheet titled "Foundation Exploration." If you're dealing with a project that doesn't have a boring log, you're stepping into the unknown, and maybe a nasty surprise.

In some cases it may be practical to do your own foundation exploration. Take a backhoe to the site and dig your own holes.

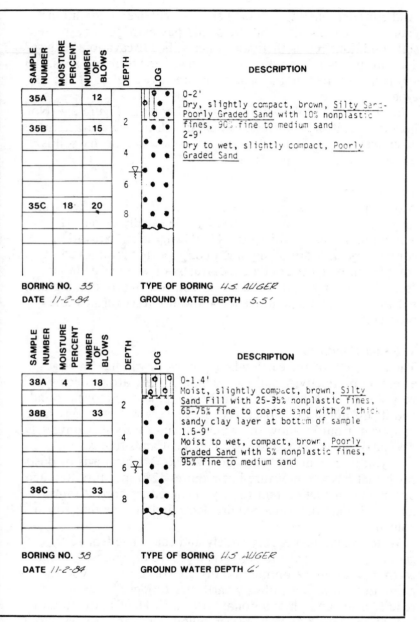

Typical boring log entries
Figure 9-18

And consider whether or not there's a "change of conditions" clause in the contract. Will the owner pay extra for unexpected soil conditions. Or is there an "unclassified excavation" clause? That means there's no difference allowed for different soils. All soil is unclassified and is paid at one rate, regardless of the soil type.

Study the table of swell factors in Chapter 2, Figure 2-4. Note how most soils have different swell and shrinkage characteristics when wet than they do when dry. And how wet or dry soil is depends mostly on the weather.

The Weather

A downpour can reduce production rates to a crawl. Everything turns into a muddy, mired mess. You're usually better off waiting until the site dries out. That's why most excavation contractors are wary of signing a contract that requires completion by a certain date, regardless of weather or other conditions over which you have no control. Before bidding that *must do* job, add a contingency allowance to cover possible problems.

The Soil Condition

The condition of the earth where you work will have a large effect on productivity and profitability. There are problems with very wet earth: it'll be hard for the equipment to move around, and it just doesn't compact well. If it's very wet, you may have to spread it out to allow the wind and sun to dry it. Conversely, dry dusty earth will require wetting. That adds to costs.

If you're moving and compacting 10,000 cubic yards of dusty earth that has 4% moisture but requires 12% moisture to reach optimum compaction moisture, you can either hope it rains a lot or calculate the water requirements and include that cost in your bid.

Water requirements vary widely and can range from 70 gallons per cubic yard to zero when you're dealing with damp earth near or above optimum moisture. If you know you're going to be working with dry earth, try to identify the water requirements and place a dollar cost on it. Here's how to do it.

Water weighs 8.34 pounds per gallon. Dry sand weighs about 2,880 pounds per bank cubic yard. Let's assume that the dry sand we have to compact contains 4% moisture; that's 2,880 x 0.04, or 115 pounds of water per cubic yard. Then divide 115 pounds by 8.34 pounds: It contains 13.8 gallons per bank cubic yard. Since 12% is the optimum moisture, how much water do you need to add to make that dry sand compact? Compacted sand weighs 3,240 pounds per cubic yard. If 12% of that is water, it holds 389 pounds of water (3,240 times 0.12 is 388.8). Now divide 389 by 8.34 to find that it holds 46.6 gallons of water.

Since your sand contains an estimated 13.8 gallons of water, just subtract that from the optimum 46.6 gallons to find how much water you need to add for compaction: 32.8 gallons per bank cubic yard.

This isn't a very precise calculation, but at least you've attempted to define a cost and allowed for it. If it takes 10 days to move 10,000 bank cubic yards and you need 100,000 gallons of water for compaction, you can add the cost per bank cubic yard to the other costs and be reasonably sure you're covered. If a watering truck costs $300 a day to rent, that's another $0.30 a bank cubic yard for watering.

Clear and Grub
Somewhere in the specifications that cover clearing and grubbing you'll find language that requires you to remove all vegetation, roots and organic matter from the site. Depending on the site, that may mean large trees, brush or grass and weeds. You problem is knowing how deep those roots go into the soil. Shallow-rooted grasses may be skimmed off by removing 2 inches of the surface. Deeper-rooted vegetation may require that nearly a foot be removed. This makes a large difference in the quantities involved.

All of these variables affect the dollar cost per unit: complexity of the take-off, swell and shrinkage factors, the variables of clear and grub quantities, the weather, water requirements and possibly drying requirements. You'll also have to consider the equipment selection, crew selection, control and motivation. Little wonder that most excavation estimators have gray hair!

Since you usually can't qualify a bid, you have to put a firm cost beside each bid item and live with the consequences.

10

EARTH MOVING EQUIPMENT AND PROCEDURES

In some ways excavation work is simpler than the pipe laying work we've been talking about. A pipe laying operation often means working large equipment on a confined city street. That requires a very well-coordinated team effort to get the pipe in the ground accurately. In excavation work, you're usually dealing with the top few feet of the ground surface, and you're dealing with it in an open area.

The risks and variables in a pipe laying job include deep trenches, groundwater, and rock, as well as confined work areas. You may face similar difficulties in earth moving, but there's usually plenty of room to work and get at the problems. The variables that make excavation risky include the weather and the soil's swell and shrinkage rates, which change the cut and fill quantities.

In simple excavation work, it's generally pretty easy to obtain good production rates — *if* you select the right equipment. We'll discuss the equipment choices in this chapter. Since dozers are so important, we'll take a detailed look at them. Then we'll talk about estimating unit production costs and walk through several typical earth moving problems.

Equipment Selection

All earth moving involves four basic tasks: load, haul, dump, and return. Of these, loading and hauling are usually the most important. Some equipment is better at loading than hauling. Some equipment is best at hauling. A few types of equipment are good at both loading and hauling. Here are some examples.

Courtesy: John Deere

Self-loading scraper
Figure 10-1

A truck can't load soil. It's used purely for hauling. A backhoe would be a poor choice for hauling. It's best at loading. Figure 10-1 shows a self-loading scraper that both loads and hauls. It uses chain-driven paddles to load the bowl and then hauls the soil to where it's needed. The bowl opens to spread the load along the line of travel.

Loaders are intended primarily for loading trucks from a stockpile. Both wheeled and tracked loaders (Figures 10-2 and 10-3) are available. Wheeled loaders are a better choice where some hauling is needed. Tracked loaders are more productive where loading is the primary task and haul distances are short.

A dozer can also be considered a self-loading hauler even though it shoves rather than carries its load. A little later in the chapter we'll look at dozers in detail.

Courtesy: John Deere

Large wheeled loader
Figure 10-2

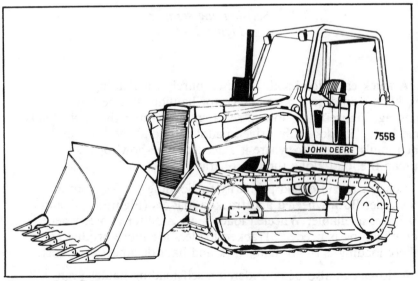

Courtesy: John Deere

Tracked loader
Figure 10-3

Push Scrapers

Push scrapers are used on extremely large earth moving projects. The push scraper loads while it's traveling forward, being shoved by a large dozer. Push scrapers are normally equipped with push blocks.

Any self-loading scraper will be more productive, especially in hard or muddy soil, when assisted by a dozer pushing on the push blocks to the rear of the bowl. But the dozer operator has to be careful. There's a danger of cutting a scraper tire with the dozer blade when the scraper turns. The dozer operator should always back off when the scraper isn't directly in line with the dozer. If the scraper doesn't have push blocks, pushing with a dozer can cause major tire damage.

Using Graders

Road graders can work to very close tolerances. They're primarily used for trimming the cut and fill area and processing material. There are two kinds of processing: Mixing and stirring material to evenly spread the water that's added to it, and spreading out material that's too wet, so it will dry.

Finish blading is the hardest part of a grader's work. A competent finish blade operator is one of the elite of equipment operators. It's his job to cut and trim earth to the tolerance of a few hundredths of a foot. When a road base or site is approaching the correct grade level, engineers place hubs at the exact finish elevation. These hubs are usually colored blue and are often called *blue tops*. Obviously, these have to stay in place during finish grading. The grader operator has to cut just to the top of the hub and no further. Hitting the hub with the blade would move the stake and require resetting. It's amazing how some operators can work right to the blue tops without hitting them.

Graders can be used for light clearing and grubbing, for maintaining haul roads and, of course, for finish grading.

Operating a Dozer

Dozers range in size from light (40 hp) utility dozers to very large machines weighing nearly 200,000 pounds that can move over 30 cubic yards in front of the blade. Other dozer work includes logging, piling brush, and clearing land. Heavyweight dozer/ripper combinations can rip and remove fractured rock, sometimes making blasting unnecessary.

Over 40 models of crawler dozers are offered by J. I. Case, Caterpillar, Deere & Co., Fiat Atllis, International-Hough, Komatsu, Liebherr-American and Terex Corp. Caterpillar and Komatsu offer very large machines with over 600 flywheel horsepower.

Each size and type of dozer has a range of appropriate uses and applications. The most common dozer models are those in the light utility and small production dozer class.

Light utility dozers— Light utility dozers have 40 to 80 horsepower and a blade capacity of 1.25 to 2.25 cubic yards. Applications include residential and commercial foundations, fine grading, trench backfill, landscaping, light logging, and swamp dozing.

Mid-sized production dozers— Mid-sized production dozers of 80 to 200 horsepower have a blade capacity of 2.75 to 5 cubic yards. Applications include trench backfill, logging, road building, land leveling, light duty ripping, tree and stump removal, and short distance production dozing. Figure 10-4 shows a mid-sized angle dozer.

Heavyweight dozers— Heavyweight dozers (200 to 700 horsepower) have blade capacities up to 25 cubic yards. They're primarily production dozers but are also widely used for ripping, push scraper work, and road building.

Dozer Operations
The basic concept behind dozer operation is simple. The machine pushes dirt in front of a blade. But productive dozing requires more than just pushing dirt around. Cutting and moving dirt efficiently is an art. If you don't lift, tilt and angle the blade just right, you'll lose most of the load before you get where you're going.

Crawler tracks convert engine power and machine weight into pushing power more efficiently than rubber tires could on dirt surfaces. And the width and length of the tracks distribute machine weight over a larger ground area. That lets the dozer

Courtesy: John Deere

A mid-sized dozer
Figure 10-4

work in soft soil where a wheeled machine would bog down.

In this section I'll explain the keys to effective dozing. We'll start with a look at the equipment itself, especially the transmissions. Then we'll examine the dozer skills you'll be using most often: basic dozing technique, dozing spoil piles, angle dozing, forestry work, and ripping.

Dozer Transmissions

The most common dozers on small earth moving jobs range between 50 and 150 horsepower. Most are equipped with transmissions that give a range of travel from 1.5 mph to 7 mph in forward and reverse at full throttle. The first crawler tractors were equipped with a direct drive transmission similar to the stick shift transmission in a car. The direct drive transmission has been largely replaced by more sophisticated transmissions.

The main difference among modern transmissions is the steering method. To steer a tracked machine, you must slow or stop one track. This lets the other track overtake the slowed or stopped track, slewing the machine to one side. Let's look at the three types of dozer transmissions.

Clutch and brake steering system combined with torque convertor and power shift transmission— This was the first steering system in crawler dozers. Steering is done by interrupting power to one track. Simply clutching the drive to one track will cause the machine to turn slowly. Interrupting power and applying the brakes to one track causes the crawler to turn more quickly.

There's a disadvantage to this type of transmission. While the machine is turning, you have pushing power from only one track. Pushing power is reduced because the moving track has to overcome the resistance of the stopped track. When using the clutch and brake steering system, try to work straight ahead. Avoid turning when loaded. Where possible, limit your turning and maneuvering to reverse passes and before picking up loads.

The steering controls of this type of dozer are usually two steering clutches (one for each track), two brake pedals (one for each track) and a master brake which applies brakes to both tracks. In some models, the clutch and brakes are combined in two steering levers. Pulling back on a steering lever first clutches the drive to the track. Further pressure applies the brakes to the track.

The *torque convertor* transmission operates on a simple principle. Imagine that you have two fans, one powered and one free spinning. If you were to place the powered fan in front of the free-spinning fan, the blasts of air from the moving fan would move the idle fan. In the case of a torque convertor, oil instead of air transfers power from the engine's flywheel to the crawler's transmission. There's no direct physical contact between the two impellers of the torque convertor.

The beauty of the design is that the spinning oil bath develops and retains power and momentum of its own. Careful design of the impellers causes the torque applied to the transmission to be greater than the torque put out by the engine.

In the fans we talked about, the air would bounce away after hitting the idle fan. In a torque convertor, the oil contributes force after having initially hit the impeller. This is *torque multiplication.* And there's another advantage to the torque convertor. The fluid coupling relieves stress during stops and starts. A conventional clutch has direct physical contact between the clutch plates. It's more subject to wear under the stress stops and starts.

In a torque convertor, the torque from the engine is multiplied and controlled by the torque convertor. The power shift transmission further controls the power and multiplies torque by means of gear ratios. Then it transfers the power to the two steering clutches and finally to the final drives. The result is low rpm torque to drive the tracks.

Torque convertor power shift transmissions with two-speed steering— Case and International-Hough make models with this steering system. The system has gear ratios in the main transmission and in the drive to the tracks. There are several travel speeds in forward and reverse. In addition, you can drive either track in high or low gear.

The advantage of this system is that it allows turns under power — both tracks are pushing, but the dozer turns toward the track that's pushing slower than the other. On Case models 1150 and 1450, you can even put one track in reverse while the other track is traveling forward. This gives the machine exceptional turning ability. Case crawlers can also be steered in the conventional manner, clutching and braking one track.

Hydrostatic drive— This is the newest crawler drive system. Hydrostatic pumps deliver hydraulic fluid under pressure to motors powering the crawler's tracks. Travel speed is controlled by the rate of flow to the motors. The drive motors respond almost instantly to changes in flow. It's relatively simple to slow or reverse one hydraulic motor and maintain forward drive in the other.

Since the speed is controlled by regulating the flow of fluid to the motor, you operate the machine with the engine at a constant speed. This is an important point. If you try to control your response time by changing engine speed, you'll stall the equipment.

Hydrostatic drive is one of the easiest systems to operate and maneuver. The machines are easy to control and capable of infinitely variable speed up to the maximum travel speed. There's no lost time shifting gears. On steep hills, in soft ground conditions and confined work areas, this system has advantages over torque convertor transmissions.

A possible disadvantage is that hydrostatic drive is less efficient in converting engine power to pushing power.

Liebherr currently offers five machines with hydrostatic transmissions. Deere and Caterpillar offer some hydrostatic drive crawler dozer models. Liebherr uses a joystick direction control. Deere uses either foot pedals or hand lever controls.

Dozing Skills

Blade position and depth of cut have the biggest influence on dozer efficiency. Other variables include direction of the dozing pass, gear selection, travel speed, and slope on which the machine is working. We'll consider each of these one at a time.

Blade position— There are two common blade positions in dozing. One is the *bulldozing* position, with the blade set at 90 degrees to the line of travel. The other is the *angle dozing* position, with the blade angled to one side.

Most small dozers are equipped with full hydraulic angling and blade tilt controls. A common dozer blade control is a four-way joystick for controlling blade lift and tilt, with a separate control lever for blade angling.

Depth of cut— The depth of cut varies with what you're trying to do and the machine size. In either case, depth of cut shouldn't exceed what the machine can handle. Too shallow a cut doesn't move enough soil. Too deep a cut causes slipping tracks and inefficient use of power.

Direction of dozing pass— Wherever possible, align the direction of dozer travel with the blade dump area before beginning a pass. Of course, some dozers can turn a load easier than others. Even so, turning will cause loss of part of the load. Don't do it unless it's absolutely necessary.

Gear selection and travel speed— Gear selection and travel speed are determined by engine horsepower, blade load, soil type, and grade. If you have the right machine for the job, you should have no trouble getting maximum load and travel speed.

You can't control the soil type and grade, but you can plan the job most efficiently around them.

The most common mistake is operating a dozer too fast during the initial phase of a dozing pass. The most critical phase of the work cycle is when the dozer is traveling forward, picking up a blade load. Take your time and cut your material smoothly and accurately. If you hurry across a bumpy or uneven area, your blade will create a washboard surface that's likely to get worse with each pass. Travel slowly when you load your blade. Remove the dirt in small slices. Speed comes later when you're transporting the blade load and making the reverse pass.

A travel speed around 1.5 mph is ideal for initial blade loading. Although dozing speeds can be as high as 5 to 6 mph, the 1.5 to 4 mph range is normally the practical maximum.

Gear changes take a little time but reduce travel time. It's usually better to change gears in the middle of the dozing pass rather than start off in a gear that's too high or travel slower than necessary. There's no set rule on which gears are best or what travel speed is right for a given application. Every job is unique.

Here's the rule of thumb I use: The motor should be turning smoothly and not lugging. An engine is lugging when it's losing speed even when you're trying to accelerate. Lugging causes excessive wear and overheating. Typically a diesel engine will turn at about 2,000 rpm. A 10-20 percent drop in rpm is a comfortable loading. More than that and the engine is lugging. Select a lower gear.

For fine grading where blade adjustments are made constantly during a dozing pass, use the maximum speed that's practical for the travel portion of the dozing cycle.

Use slow speed to load the blade. Increase speed for transporting the blade load.

Reverse travel offers another opportunity to minimize cycle times. As the dozer approaches the beginning of the next pass, steer it into position to make the pass without further steering adjustment. This lets you concentrate on speed control and blade adjustment as you begin your next dozing pass.

As you work an area, you'll leave rows of spoil as part of the load is lost off the blade. These rows are called *windrows.* Crossing a windrow too fast can bounce the operator around in the cab. That can be dangerous. While I stress speed and efficiency, don't forget safety. Be reasonable.

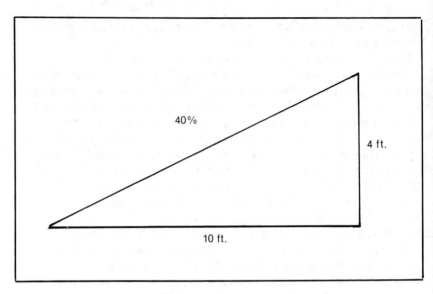

40% downgrade doubles dozer's blade load
Figure 10-5

Downgrade dozing— In earth moving, it's common to use the percent of grade method to describe a slope. For example, a slope which rises 50 feet in 100 feet of horizontal distance is a 50 percent slope. (50 divided by 100 times 100 equals 50 percent.) A 100 percent slope is the same as a 45-degree angle. It rises 30 feet in 30 horizontal feet. Figure 10-5 shows a 40 percent grade.

A positive grade is an uphill slope. A negative grade is a downhill slope. Downgrade dozing increases the dozer's blade capacity and adds to the pushing force available. Gravity helps in both cases. The advantage is dramatic. *The blade load capacity of a dozer doubles if the load is pushed down a negative 40 percent grade.* The model 850 dozer with its 2.5 cubic yard blade can push 5 loose cubic yards down a 40 percent slope.

Dozing uphill is just the opposite. A positive grade has a negative effect on dozer production. When a dozer has to push its blade load up a 40 percent slope, the blade capacity is cut in half.

How many times have you seen dozers struggling to backfill large sewer trenches by working from the base of the spoil pile?

Losing the blade load
Figure 10-6

A better way is to make an access ramp and level out the top of the pile. Then the operator can push spoil downgrade to the trench. That will increase production dramatically.

A dozer blade tends to lose its load as the dozing distance increases. The blade load is triangular. As the machine moves forward, the load is gradually lost to each side and under the blade. Eventually there's nothing left of the original load. Figure 10-6 shows how a dozer loses its blade load.

You can reduce the amount of load lost by *slot dozing*. The dozer works in a slot about 2 feet deep and the width of the blade. Pushing soil in the slot keeps the load from slipping away. See Figure 10-7. You can get nearly the same result by having two dozers work in tandem, pushing the same load side by side. Another alternative is to use a dozer equipped with a U-shaped blade when pushing soil longer distances. The U-blade is more efficient for dozing long distances. These blades can be fitted to many large dozers.

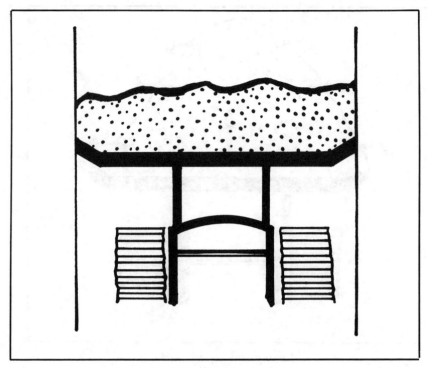

Slot dozing
Figure 10-7

Dozing spoil piles— When you're spreading spoil, plan the work carefully. There are good ways and bad ways to spread a pile of earth. The key is getting maximum capacity from each pass. Figure 10-8 shows how to break a spoil pile down to a manageable size. Follow the sequence of cuts numbered from 1 through 5 in the drawing.

Angle dozing— Most dozers can angle the dozer blade. On small utility dozers, you may be able to angle the blade hydraulically. On other machines you have to adjust the angle manually.

   In either case, blade angling increases the versatility of the machine. An angled blade can *sidecast* material and position it at 90 degrees to the line of travel. This is important, since it's often impossible to position the tractor directly behind the material you need to bulldoze. Angle dozing lets you move soil when there's no way to attack it from the rear.

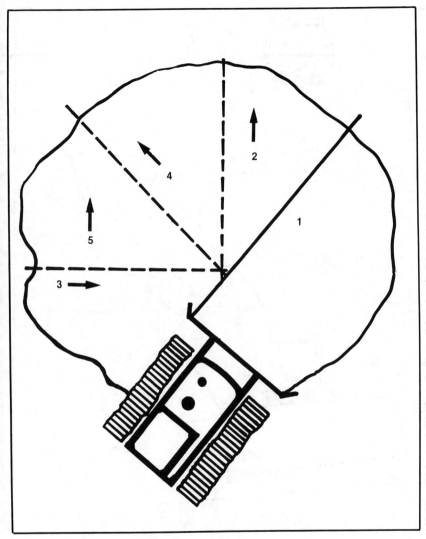

Breaking down a spoil pile
Figure 10-8

To sidecast material, use the blade tilt control in combination with the angle control. Figure 10-9 shows how to do this. In section A, the right side of the blade is tilted down and angled forward. As the forward edge of the blade slices into the pile, resistance of the soil tends to pull the blade deeper into the cut.

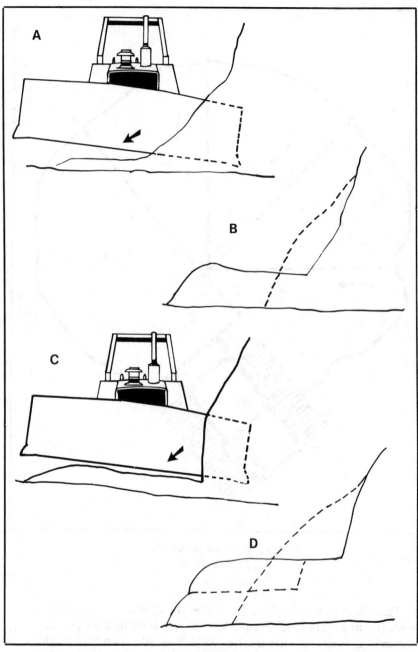

Sidecasting
Figure 10-9

Angled blade causing machine
to slew into bank
Figure 10-10

It's easy to make a cut that's too deep. The angled blade tends
to slew the machine into the bank, as shown in Figure 10-10.

On a small slope, you'll start sidecasting at the bottom of the
slope and work up. Section A in Figure 10-9 shows the first cut.
In section B, the first cut has been finished. Notice that the
spoil has been pushed out of the cut to form a ramp at the left.
In section C, the dozer is making the second pass a little higher
up the slope, again depositing spoil to the left. In section D, the
second pass has been made, preparing the ramp for the third
pass.

Where you're cutting right-of-way across a steep slope, it's
faster to reverse the process, starting at the top. If you started
at the bottom, spoil would fall away too far, leaving the
machine without a ramp to work from. If you can't build
enough ramp with each pass when working from the bottom,
start cutting from the top of the slope.

Reading Slope Stakes

In other parts of this book I've spoken of slopes, grade, gradient, and "fall." In earth moving, a slope normally refers to the side slope of an excavation. We lay pipelines to a percent or decimal grade. A road's grade is usually expressed in percent of grade: a 5 percent downgrade sign along a highway tells you that you're approaching a section of road that descends 5 feet in each 100 feet of horizontal travel.

The slope of a hillside is usually expressed in percent. The slope of a cut or trench wall is usually expressed as a ratio. A 1:1 slope is a slope of 1 foot horizontally in each 1 foot of vertical measure. A 1/2:1 slope means 1/2-foot horizontal rise in 1 foot vertical distance. A slope of 1/2:1 is about the practical minimum for a stable slope in soil. Usually the side slope will be more gradual, perhaps 3:1 or even 5:1.

Something to Remember

The dozer operator is responsible for the safety of those working around the dozer. Nothing on the job is worth as much as the life of someone on your pipeline crew. Don't compromise safety for better production. When you consider the mass of soil and rock or the size of logs a dozer can move, the human body is a frail object. Don't forget that. Use common sense.

When working in the woods, keep the work area clear of trimmed branches so as not to obstruct the operator's vision. Be especially alert when backing.

Tracked equipment can work on steep ground, especially if the dozer is facing directly up or down a slope. But be cautious on side slopes, especially if the surface is wet or spongy. Rollovers can be fatal.

If you're inexperienced at working dense woods, be especially careful. Brush can conceal old stumps that will drastically alter a dozer's center of gravity with little warning. Small poles and branches have a way of getting into the operator's platform, injuring the operator and interfering with the machine's controls.

When using a winch in your logging work, always line the machine up in the direction of pull before starting the winch. Use a cable with a breaking strain greater than the winch capacity. And never start winching until the ground workers give you an "all clear" signal.

Working on Steep Slopes

A slope over 60 percent is usually called a *knee shaker*. A dozer can remain relatively stable on slopes up to 100 percent (1 foot of rise in each 1 foot of horizontal distance), but you'll rarely find a smooth and even 100 percent slope on natural ground. Usually the slope will be steeper in some places and shallower in others. A 100 percent slope may have sections that are well over 100 percent.

Most dozers will climb a 100 percent slope. The problem will be in maneuvering, turning and stopping. A crawler equipped with a conventional transmission is the hardest to control on a steep slope. To turn left while traveling forward requires the left track to be slowed or braked. When the transmission is clutched, the dozer will stop. Unless you immediately apply the brakes, the machine will begin to travel backward down the slope. If the left brake is still applied, the front will slew rapidly to the right. Here's the control sequence to use to avoid this: throttle down, apply the main brake, then release the left brake as the clutch is pulled in.

Maneuvering a dozer over the lip of a dropoff can also be dangerous. If it's done carelessly, the operator will be thrown around in the cab, probably losing control and maybe suffering some injury. It's important to reach the point of balance very slowly, then ease ahead until the dozer slowly falls into position on the slope.

Ripping

Large dozers with engines of several hundred horsepower can rip rock that's nearly solid. The standard utility dozer is too small to be effective in real rock work. But many small dozers are fitted with rippers that can loosen hard soils and relatively soft sedimentary or decomposed rock.

There's an art to breaking out rock. Most rock formations and soils were originally laid in horizontal beds. Movement in the earth's crust upset these flat layers and jumbled them in every direction — even edgewise and at differing angles to horizontal. If you can't penetrate hard ground from one direction, try approaching it from a different angle. The rock may be less resistant when hit from a different direction.

Larger dozers can exert a force of 20 to 30 tons on a ripper tip. That's enough to rip and fragment most rock. Ripper penetration depends on the angle of the ripper. A tip angle of 60 degrees allows good initial penetration. Try 45 degrees to exert good lifting or prying force. To pry out boulders, use an angle of 35 degrees while raising the shanks.

The larger machines designed for heavy duty ripping are usually fitted with adjustable rippers. Smaller dozers are fitted with fixed shank rippers useful for loosening hard earth prior to blade work. There are two very important points to remember when using rippers. First, always be prepared to back off. If you hit an immovable object, press the clutch immediately to avoid damage. And second, never turn the dozer while the rippers are in the ground.

Choosing the Right Equipment

Having the right equipment for earth moving is as important as it is for pipe laying. You wouldn't choose a self-loading scraper to excavate a downtown basement, for example. While a dozer can move earth several hundred feet quite efficiently, it loses its efficiency after a certain point. A better choice for a longer distance is a self-loading scraper or a loader and trucks. A combination of different equipment can also be efficient in some circumstances. A dozer stockpiling for easy loading by wheeled loaders or tracked excavators may be a cost-effective combination.

As with all production estimating, it's hard to predict production rates precisely. While you may not be able to forecast exact rates, you can avoid predicting the impossible.

Estimating Production

To haul, dump and return over a haul distance of 2,000 feet in 3 minutes, the truck would have to average over 30 mph and dump the load in 1 minute. Remember, we're talking an average speed here. And that's just not going to happen. In city traffic, the truck will probably be lucky to average 5 mph. At this average speed, it'll take over 9 minutes to travel both ways plus the time it takes to dump. If you based your production estimate on 3-minute cycles, you'd lose your shirt. But look on the bright side — if you survived the experience, you'd be a more experienced estimator!

Production estimates require mathematical calculations, previous production records, data gathered by observation, common sense, and bid tabulations of similar projects. But don't depend on any of these sources alone. Use them in combination. If you hauled earth for $2 a loose cubic yard last year, you can probably do almost as well this year, or maybe even better. But you're not going to make a profit at $1 a cubic yard on a similar job. Compare apples to apples.

Bid tabulations (bid abstracts) can indicate the going rates for earthwork. But they can also give a distorted picture. Some estimators purposely distort the unit prices they submit on public projects. Here's why. If I consistently bid earth moving work at $1.50 a cubic yard, it won't take Joe's Backhoe Service down the street long to figure out that bidding $1.40 a cubic yard would cut me out of the project.

Bid tabs can be useful. But their value is limited. Your bids have to be based on your own costs and production rates.

Cycle Times
Figures 10-11 and 10-12 show travel time for short and long hauls. Remember, a haul is a two-way process. Hauling empty usually goes faster than hauling loaded. Every haul cycle has four parts:

1) Load
2) Haul out
3) Dump
4) Return

Add to this cycle the unproductive time, like standing time, waiting times, breakdown time, service time and the all-important *load factor.*

Loader bucket fill factors— These factors correct the rated bucket capacity for particular types of soils. For example, a loader equipped with a bucket rated at 3 loose cubic yards may actually be able to hold 110 percent of its rated capacity of

Distance (in feet)

Speed (in mph)	100	150	200	250	300	400	500	600	1,000	5,280
1	1.14	1.70	2.27	2.84	3.41	4.55	5.68	6.82	11.36	60.00
2	.57	.85	1.14	1.42	1.70	2.27	2.84	3.41	5.68	30.00
3	.38	.57	.76	.95	1.14	1.52	1.89	2.27	3.79	20.00
4	.28	.43	.57	.71	.85	1.14	1.42	1.70	2.84	15.00
5	.23	.34	.45	.57	.68	.91	1.14	1.36	2.27	12.00
6	.19	.28	.38	.47	.57	.76	.95	1.14	1.89	10.00
7	.16	.24	.32	.41	.49	.65	.81	.97	1.62	8.57
8	.14	.21	.28	.36	.43	.57	.71	.85	1.42	7.50
10	.11	.17	.23	.28	.34	.45	.57	.68	1.14	6.00
12	.09	.14	.19	.24	.28	.38	.47	.57	.95	5.00
15	.08	.11	.15	.19	.23	.30	.38	.45	.76	4.00
20	.06	.09	.11	.14	.17	.23	.28	.34	.57	3.00
25	.05	.07	.09	.11	.14	.18	.23	.27	.45	2.40

Travel time in minutes - short haul
Figure 10-11

Speed (in mph)	Distance (in feet)									
	5,280 (1 mile)	7,920 (1.5 mile)	10,560 (2 mile)	15,840 (3 mile)	21,120 (4 mile)	26,400 (5 mile)	52,800 (10 mile)	79,200 (15 mile)	105,600 (20 mile)	132,000 (25 mile)
20	3.00	4.50	6.00	9.00	12.00	15.00	30.00	45.00	60.00	75.00
25	2.40	3.60	4.80	7.20	9.60	12.00	24.00	36.00	48.00	60.00
30	2.00	3.00	4.00	6.00	8.00	10.00	20.00	30.00	40.00	50.00
35	1.71	2.57	3.43	5.14	6.86	8.57	17.14	25.71	34.29	42.86
45	1.33	2.00	2.67	4.00	5.33	6.67	13.33	20.00	26.67	33.33
50	1.20	1.80	2.40	3.60	4.80	6.00	12.00	18.00	24.00	30.00
55	1.09	1.64	2.18	3.27	4.36	5.45	10.91	16.36	21.82	27.27
60	1.00	1.50	2.00	3.00	4.00	5.00	10.00	15.00	20.00	25.00

Travel time in minutes - long haul
Figure 10-12

moist loam. That's a fill factor of 1.1. But when loading poorly shot rock containing large irregular boulders, its actual bucket load may amount to only 50 percent of its rated bucket capacity. To use the bucket load correction factor, multiply the rated bucket load by the appropriate material load factor.

Damp sand, moist loams: 1.1 times rated capacity
Sand and gravel mixture: 1 or 1.05
Clean gravels (pea gravel):   0.85 or 0.95
Rock-earth mixture: 0.9
Well-blasted rock or cobbles 6''-10'' diameter: 0.75 or 0.85
Poorly blasted rock (slabs and boulders):   0.5 or 0.6

Truck Capacity

It's easy to calculate the capacity of a truck. The bed of a truck is a rectangle. The formula for the volume of a rectangle is height times width times length. If the truck bed measurements are in feet, divide your answer by 27 to convert to loose cubic yards.

Here's a sample calculation:

$$\frac{H \times W \times L}{27} = \text{volume in L.CY}$$

If you stand on a freeway overpass, you'll probably see partially loaded dump trucks. Maximum loads are mandated by law. Spillage of rocks and soil on a highway is both illegal and dangerous. But spillage on the site is seldom a problem. If your trucks don't leave the site, fill them to overflow. All too often, a truck that could be filled completely is left partially unfilled. That's caused by inattention on the part of operators and drivers.

To the experienced excavator, whether he specializes in pipework, earthwork, or both, the job site is a study in maximum efficiency. Every effort should be made to reduce the *cost per unit*.

Cost Per Unit Tells the Story

Most excavation and pipeline contractors admire efficient, highly productive heavy equipment. The giant machines growl, revolve, cycle and produce in a repetitive orchestra of activity. It's a spectacle to behold. It's also a demonstration of human efficiency and inefficiency. Every job site has its share of both.

But the utility of the machines is measured in cost per unit. The unit may be time: seconds, minutes or fractions of an hour, days, weeks or months. It may be cost per loose cubic yard or per linear foot. In either case, cost per unit determines the profit or loss — and ultimately your success or failure. But even if you're not interested in making money, there's no reason to waste time, horsepower, human power, equipment power or energy.

Efficiency Creates Profits

To the earth mover, efficiency means profits. Even relatively minor inefficiencies can turn a job into a losing proposition. With good job management and a minimum of inefficiency, the same job can return an excellent profit. To stay in business, you should see each equipment cycle in cost per unit.

If a truck is moving 60 cubic yards per hour at a cost of $60 per hour, the cost per yard is $1.00. If the same unit only moves 55 cubic yards per hour, that's a cost of $1.09 per yard. At 50 cubic yards an hour, the cost becomes $1.20 per yard. If your bid was at $1.10 per yard, you're paying the owner ten cents a yard to move his dirt!

When calculating unit costs, be sure to include all your costs, including both direct and indirect overhead — and profit, of course.

Until you see your projects as a *unit cost per component,* you're not likely to succeed as a pipeline or excavating contractor. You'll be paying more than necessary for each unit of work completed, reducing your productivity and profit. Ignoring unit costs is like burying your head in the sand.

You can expect that most of the people who work for you aren't conscious of the need to reduce unit costs. You need to change that. With tact and patience, you have to make high productivity and lower unit costs a high priority for everyone on your payroll.

Formulating a Work Plan

Employees and equipment never work at top efficiency without a good plan. Matching equipment and people to the task, and job layout are the keys. Let's look at some examples.

Loader Production

The average cycle time for a wheeled loader ranges between 30 and 45 seconds (0.5 to 0.75 minutes)

Loading production is the number of cycles required per load multiplied by time per cycle. That gives you the time needed to load each unit.

The number of cycles required per load depends on the capacity of the haul unit and the capacity of the loader. Loading cycle time depends on the type of material being loaded, the distance the loader has to travel, and the travel speed between loading and dumping.

Here are typical loader cycle times for loading trucks from loose stockpile on a flat level surface:

Small loader (1 to 3 L. CY): 0.5 to 0.67 minutes (30 to 40 seconds) per cycle.

Mid-sized loader (3 to 5 L. CY): 0.6 to 0.75 minutes (35 to 45 seconds) per cycle

Estimating Scraper Production

In spite of its size, a scraper is a relatively simple machine to operate. The bowl or pan is raised or lowered hydraulically. The elevator lifts the spoil up and back into the bowl. To dump, a hydraulically operated gate shoves the material forward, reversing the elevator to assist in unloading. But while it may be simpler to operate than many of the heavy machines, don't assume an inexperienced operator will be able to manage. As always, a skilled operator will outproduce an inept operator every time.

A scraper's work cycle includes travel to the dump site (time depends on the haul distance), dumping (40 seconds to 1 minute), and return travel. Look at the travel time charts in Figures 10-11 and 10-12.

You'll have to add these variables to the basic cycle time:

• Maneuvering into position to load or dump. This is a particular problem in confined sites.

• Delays caused by bottlenecks where one machine is waiting for another to pass.

• Working upgrade is significantly slower than hauling on level ground.

• Fine trimming where the scraper has to make shallow cuts very accurately or where work is in cul de sacs or follows complex layouts.

• The condition of the surface over which the machine has to travel. A heavy machine will sink into a soft surface. Even if the machine doesn't get mired or stuck, the tires will have to overcome the resistance of the pliable surface. This is called *rolling resistance.* It dramatically increases cycle time because any resistance demands more horsepower at the wheels. The operator has to downshift to provide it. Aside from operator ability and uphill hauling, rolling resistance is the greatest production variable in scraper production.

Haul speeds vary from 4 to 6 mph in difficult conditions to as high as 20 to 30 mph over a well-groomed haul road. Remember that a scraper moving at 12 mph on a empty return travel will average much less than that for the entire return trip. Some time is wasted in acceleration and deceleration. The best way to estimate cycle time is to observe several jobs under varying conditions.

Once you've found the cycle time, there are several basic formulas for estimating scraper production. First, find the cycles per hour. Then multiply that by the loose cubic yards per load. That gives the loose cubic yards per hour:

$$\frac{60 \text{ minutes}}{\text{cycle time (in mins.)}} = \text{cycles per hour}$$

Cycles per hour x L.CY per cycle = L.CY per hour

Truck Production
Truck production depends on loading time, travel time, and dumping time, plus the time needed for positioning at the loading and dump sites. This is usually called *spotting* time. Driver skill has a major influence on spotting time. Typical travel times are shown Figures 10-11 and 10-12.

Trackhoe Production

Trackhoe loading times depend on the quantity per cycle and the number of cycles per unit.

$$\frac{\text{Load requirement}}{\text{L.CY per cycle}} = \text{cycles required per load}$$

Cycles required per load x time (in mins.) = loading time (in mins.)

Sample Earth Moving Problem

Suppose your job is to excavate a basement that's 100 feet square and 10 feet deep. It's in an urban area. For simplicity, we'll assume that it has vertical walls that will be shored during the building phase. First, let's find out how many bank cubic yards we have to move:

$$100 \text{ x } 100 \text{ x } 10 \div 27 = 3{,}704 \text{ B.CY}$$

If the soil is firm and dry loam, the swell factor will be 1.35:

$$3{,}704 \text{ x } 1.35 = 5{,}000 \text{ L.CY}$$

Because of the urban location, this will be a trucking operation. The first step will be to locate a dump site as near as possible to the excavation site. Then we can establish a realistic haul and return time. Clean earth is valuable. We'll look for someone who needs fill dirt within a reasonable haul distance. Failing that, we'll find a temporary stockpile or dump site.

Wherever city traffic is involved, it's best to travel the haul route to find a realistic trucking time. It may make sense to work nights or weekends to avoid traffic delays. If you don't have your own trucks, consider hiring trucks by the hour or day or by the cubic yard or trip. When paying by the load, trucking time will usually be less than when paying by the hour. People work harder when they have the motivation of getting paid by output, not hours on the job.

Loading Equipment

An area that's 100 feet by 100 feet is a pretty small space for maneuvering trucks and equipment. A ramp is needed so the trucks and equipment can get into the 10-foot hole. A tracked excavator, on the other hand, could load trucks from the existing surface grade. But let's minimize loading time by using a suitable trackhoe.

A trackhoe fitted with a 2 loose cubic yard (L. CY) bucket would load a 12-yard truck in six cycles or less. With a digging cycle of 20 seconds (0.34 minutes), trucks could be loaded in two minutes. Allowing two minutes for the trucks to position themselves for loading, we could load a truck every four minutes.

If we have a haul, dump, return cycle of 45 minutes for the trucks, we could in theory use 10 or 11 trucks. The danger is that we would create a traffic jam of trucks lining up to be loaded out. The chances are that some trucks would be delayed, others would make faster trips. The result would be expensive equipment held up and costing you money.

If we use eight trucks, we could load then all in 32 minutes. If the cycle time for the each truck is 45 minutes, the trackhoe will be waiting for trucks 13 minutes out of every 45. Is that good or bad? Should we have trucks waiting for the trackhoe or the trackhoe waiting for trucks? Why not schedule exactly as many trucks as the trackhoe can handle? Let's compare costs and see whether we want trucks or the hoe waiting.

Assume that trucks with a driver cost $40 per hour each or $320 per hour for the eight. The tracked excavator will cost about $60 per hour. If we're going to lose efficiency somewhere, it's better to lose it off the trackhoe rather than have the more expensive trucks held up. You can see that it's better to schedule too few trucks and have the trackhoe waiting than it is to have too many trucks.

Trucking Costs

At $40 per hour, the eight trucks will cost $320 per hour times eight hours, or $2,560 per day. Assuming a 45-minute haul cycle, each truck could haul 10 loads per day, 12 yards per load. That's 120 yards per truck, or 960 loose cubic yards total for the eight trucks each day.

Since we're dealing with 5,000 L. CY, in theory we could expect to excavate and haul out this material in five working days. But before we make a firm commitment based on these calculations, let's check our premises and assumptions.

1)   Are we using the right equipment for the job?
Scrapers, loaders, and dozers are not good choices. This reaffirms the choice of a tracked excavator and trucks.

Are 12-yard trucks the best choice? They can be fitted with trailers that will haul an additional 10 L. CY. The loading time would increase, but the haul travel time wouldn't increase significantly. The trucks and trailers would cost more per hour, but the cost to haul could decrease because they're hauling a larger payload.

Another choice is the larger tractor trailer haulers called "belly dumps." These have a capacity of around 20 cubic yards. Again, they cost more on an hourly basis but haul a larger payload. For an urban lot, the final choice would be dictated by site access. There's little point in selecting a cost-effective haul unit if it can't get into and out of the site.

In this case, we like our first choice of equipment. We'll base our cost estimate on 12-yard dump trucks and a trackhoe to load them.

2)   The next decision is the projected production rate and cost. At optimum production of 960 L. CY, it will take 5.2 days:

$$5,000 \text{ L. CY} \div 960 = 5.2 \text{ days}$$

Unfortunately, we can't expect optimum production. To allow for lost time due to delays, we'll reduce our optimum production rate by the 0.83 efficiency factor we discussed earlier in this book.

$$960 \times .83 = 797 \text{ L. CY per day}$$
$$5,000 \text{ L. CY} \div 797 = 6.27 \text{ days}$$

Since it's always better to err on the side of caution, we'll round this up to seven working days.

Our trucks will cost $2,560.00 each working day. The trackhoe costs $480 a day. That's a daily equipment cost of

$3,040 — plus labor, any auxiliary equipment, supervision, mobilization, and our overhead and profit.

Seven days at $3,040 gives us $21,280.00 for our total equipment cost. Dividing that by the 5,000 bank cubic yards (B. CY) we have to move, we find the projected equipment cost to be $4.26 per cubic yard for loading and hauling.

Risk Management and Contingency
We've identified the probable job cost. But there's something else to consider. If a reliable trucking company will do the hauling for $3.25 L. CY or $39 per load, would you accept their bid? Of course you would! You'd be saving over $1 per cubic yard — and the risk of loss would be reduced considerably. Giving the work to a subcontractor fixes your cost and reduces your risk. If you do the trucking yourself, the actual cost could be much higher than your $4.26 estimate.

On work that isn't subcontracted, most pipeline and excavation contractors add a few percent to cover contingencies. A contingency allowance covers costs that can't be forecast before work begins. Excavation is seldom done faster than planned. Most surprises will increase the cost, not decrease it. The right amount to add for contingencies such as overtime, low productivity or poor scheduling depends on the contractor and the job. For most excavation, 2% to 5% may be enough. If the plans are poorly drawn or unclear, if the water table is high, if bad weather is expected to reduce productivity, or if any of hundreds of other variables can't be accounted for, the contingency allowance may have to be much higher.

You'll never have the luxury of an absolutely certain profit. Instead, try to earn a fair profit if the job goes as predicted and a higher profit if production is higher than expected. If unexpected problems come up, there should be enough contingency in the job to cover most of the extra costs.

Here are some of the risks your contingency allowance may have to cover:

Clearing and grubbing— The clear and grub specification is usually worded something like this: *Clearing: All vegetation including snags, brush, rubbish and other vegetation shall be removed. . . Grubbing: Removal and disposal of stumps, roots, matted roots. Organic or metallic debris shall be removed to a depth of 18''. . .*

In other words, you're required to remove all the materials described, no matter whether it's 100 or 5,000 cubic yards. A site covered with grass and weeds may be easy to strip in the dry season. Skimming the top 2 inches may be enough. Roots tend to separate very easily from dry soil. During the rainy season, muddy soil will cling to the roots. The risk is twofold: First, more soil to haul, and second, a good chance that you'll have to buy and import soil to replace what was removed.

Incomplete or faulty grading plans— Another example of a high risk job is where the grading plan is incomplete or of poor quality. If you're not given enough accurate data to average volumes, your estimates may be imprecise. When we're speaking of thousands of cubic yards of earth, a mistake of a few percent can increase the cost of completion dramatically. If imported fill costs $5 to $12 per cubic yard, bringing in a few hundred cubic yards can erode the profit very quickly.

The weather— The weather has a dramatic affect on production and costs. Equipment like wheeled and tracked backhoes will work in mud with very little loss in production, especially if you have experienced operators. But scrapers, trucks and graders don't work very well in mud.

Bankers understand risk. The best loan rates go to their blue chip customers. You and I pay higher rates when we borrow because there's more risk for the bank. Do your estimating the same way. If you can remove most of the risk by accepting the bid of a subcontractor, reduce or eliminate the contingency allowance.

Another Sample Problem: Cut and Fill

Let's say you're bidding the site work for a new apartment complex. Assume that you operate primarily as a pipe contractor. On this job, you'll have a better chance of getting the work if you bid utility installation and site work as a package. Many general contractors like to award site work and utility installation to the same bidder. That reduces conflicts and confusion on the site.

If you're primarily a pipe contractor, take a long, hard look before you leap. Excavation contracting has uncertainty and risks, and they're not the same risks a pipe contractor is used to dealing with. Even if you use the best information available about the soil type and its behavior (swell, shrinkage, water requirements, drying requirements) and the best estimating procedures, there's still the chance of an estimating error that will cost you money.

Remember this. The soil swell and shrinkage tables in this book are subject to about a 10 percent error. They're based on *averages*. But you can still use them to make estimating decisions. If you use sand fill with an average shrinkage of 11 percent, you're better off than if you use a silty clay fill with a shrinkage of 17 percent. You'll need less of the sand. This is especially important if you're paying for the fill material.

You can also get hurt in the conversion from tons to cubic yards. Let's assume you're pricing sand to use for fill. Say that sand supplier A sells sand for $5 a ton. Supplier B charges $6 a ton. Obviously you'll buy sand from supplier A. But wait a minute. If supplier A operates a wet pit where the sand is dredged out of water, it will weigh about 3,200 pounds per L. CY. At $5 a ton, that's $8 per L. CY. If supplier B sells dry sand weighing 2,600 pounds per L. CY, that's $7.80 per L. CY at $6 a ton.

At an extra $.20 per L. CY, the wet sand will compact better and may well be worth the extra cost. Make sure that what appears to be a savings isn't going to cost you in the long run. Always agree on the precise weight and condition of imported fill before you make the deal.

It's obvious that the degree of compaction has a dramatic effect on the quantities you'll have to move. The precise shrinkage will be hard to determine before it's placed and compacted. Another variable is the ease of compaction. Generally speaking, a course-grained soil is easier to compact than a fine-grained soil. The course-grained soils drain well so they tolerate overwetting. Fine-grained soil dries slowly and only compacts easily at the correct moisture content. Overwetting a fill material is one of the easiest ways to attain compaction — but only if it will dry quickly.

As an earth moving contractor, you're not simply moving earth at a dollar cost per L. CY. You're moving a specific type of soil with specific behavior characteristics that will require specific treatments to attain the contract requirements. A calculated fill requirement may actually run slightly less than projected when good quality uniform fill material is used. It may also run up to 30 or 40 percent more than projected, if a poor choice is made.

Balancing the Cut and Fill

The aim of most grading plans is to have the cut quantities and the fill requirements *balance.* Ideally, all material removed from the cut areas can be used in the fill areas, requiring no hauling to or from the site. That isn't always easy. Remember that soil swells when excavated and then shrinks when compacted. Swelling and shrinking aren't always the same for every type of soil.

The first rule is not to take anything for granted. The plans may include a neat quantity takeoff that looks like this:

Cut quantity: 10,000 B. CY   Fill quantity: 9,259 C. CY

How very convenient that common earth shrinks about 8% from bank cubic yards to compacted cubic yards:

$$\frac{10{,}000}{1.08} = 9{,}259$$

It will work out perfectly, right? *Don't take anything for granted!* Check it out yourself. If the contract provides payment for fill that's needed or excess that has to be hauled off, make a guess on which will be the case. Bid highest on the most likely result, either export or import. Figure 10-13 shows the quantities of material (in bank cubic yards) in a cut or fill area of 10,000 square feet. It ranges from almost 30 B. CY for 1 inch of cut or fill to over 4,000 B. CY for 12 feet of cut or fill.

If the contract doesn't provide extra payment for import or export of soil, protect yourself. Figure the quantities carefully. Bid exactly the quantity you feel would have to be imported or exported — at your cost plus a reasonable profit. I've seen major mistakes in excavation plans. It's easy to misread a scale or make a math error. Regardless of the professionalism, reputation and expertise of the designer and engineer, *do not take their quantities as fact unless they are paying for any mistake.*

	Cut	
(inches)	(feet)	B.CY of material
1"	.08	29.63
2"	.17	62.96
3"	.25	92.59
4"	.34	125.93
5"	.42	155.56
6"	.50	185.19
8"	.67	248.15
18"	1.50	555.56
	2.00	740.74
	3.00	1111.11
	4.00	1481.48
	5.00	1851.85
	6.00	2222.22
	7.00	2592.59
	10.00	3703.70
	12.00	4444.44

B.CY of material per 10,000 SF of cut or fill
Figure 10-13

Calculate the grid quantities as explained in this manual. Average the elevations. Average the proposed subgrade elevations. Does your conclusion verify the engineers quantities? This kind of averaging isn't precise, but it will uncover a gross error.

Estimating is an art, not an exact science. I like to use the term "feeling comfortable." Do you feel comfortable with your premise, your calculations and your conclusions? If so, you've probably prepared an estimate that will cover your costs and make a reasonable profit.

INDEX

Practical References for Builders

Excavation & Grading Handbook Revised

Explains how to handle all excavation, grading, compaction, paving and pipeline work: setting cut and fill stakes (with bubble and laser levels), working in rock, unsuitable material or mud, passing compaction tests, trenching around utility lines, setting grade pins and string line, removing or laying asphaltic concrete, widening roads, cutting channels, installing water, sewer, and drainage pipe. This is the completely revised edition of the popular guide used by over 25,000 excavation contractors.
384 pages, 5½ x 8, $22.75

Estimating Excavation

How to calculate the amount of dirt you'll have to move and the cost of owning and operating the machines you'll do it with. Detailed, step-by-step instructions on how to assign bid prices to each part of the job, including labor and equipment costs. Also, the best ways to set up an organized and logical estimating system, take off from contour maps, estimate quantities in irregular areas, and figure your overhead.
448 pages, 8½ x 11, $39.50

Construction Forms & Contracts

125 forms you can copy and use — or load into your computer (from the FREE disk enclosed). Then you can customize the forms to fit your company, fill them out, and print. Loads into *Word for Windows*, *Lotus 1-2-3*, *WordPerfect*, *Works*, or *Excel* programs. You'll find forms covering accounting, estimating, fieldwork, contracts, and general office. Each form comes with complete instructions on when to use it and how to fill it out. These forms were designed, tested and used by contractors, and will help keep your business organized, profitable and out of legal, accounting and collection troubles. Includes a CD-ROM for *Windows*™ and Mac. **432 pages, 8½ x 11, $41.75**

 Craftsman Book Company
6058 Corte del Cedro, P.O. Box 6500
Carlsbad, CA 92018
☎ 24 hour order line
1-800-829-8123
Fax (760) 438-0398

Order online

http://www.craftsman-book.com
Free on the Internet! Download any of Craftsman's estimating costbooks for a 30-day free trial! http://costbook.com

Name _____

e-mail address (for order tracking and special offers)

Company _____

Address _____

City/State/Zip _____ ○ This is a residence
Total enclosed_____(In California add 7.25% tax)
We pay shipping when your check covers your order in full.
In A Hurry?
Use your ○ Visa ○ MasterCard
○ Discover or ○ American Express

Card _____

Exp. date_____Initials_____

Tax Deductible: Treasury regulations make these references tax deductible when used in your work. Save the canceled check or charge card statement as your receipt.

10-Day Money Back Guarantee

○ 39.50 Basic Concrete Engineering for Builders

○ 74.00 CD Estimator — Heavy

○ 41.75 Construction Forms & Contracts with a CD-ROM for *Windows*™ and Macintosh.

○ 39.50 Estimating Excavation

○ 22.75 Excavation & Grading Handbook Revised

○ 39.00 Getting Financing & Developing Land

○ 29.00 Pipe & Excavation Contracting

○ FREE Full Color Catalog

Prices subject to change without notice

Getting Financing & Developing Land

Developing land is a major leap for most builders — yet that's where the big money is made. This book gives you the practical knowledge you need to make that leap. Learn how to prepare a market study, select a building site, obtain financing, guide your plans through approval, then control your building costs so you can ensure yourself a good profit. Includes a CD-ROM with forms, checklists, and a sample business plan you can customize and use to help you sell your idea to lenders and investors. **232 pages, 8½ x 11, $39.00**

CD Estimator — Heavy

CD Estimator — Heavy has a complete 780-page heavy construction cost estimating volume for each of the 50 states. Select the cost database for the state where the work will be done. Includes thousands of cost estimates you won't find anywhere else, and in-depth coverage of demolition, hazardous materials remediation, tunneling, site utilities, precast concrete, structural framing, heavy timber construction, membrane waterproofing, industrial windows and doors, specialty finishes, built-in commercial and industrial equipment, and HVAC and electrical systems for commercial and industrial buildings. **CD Estimator — Heavy is $74.00**

Basic Concrete Engineering for Builders

Basic concrete design principles in terms readily understood by anyone who has poured and finished site-cast structural concrete. Shows how structural engineers design concrete for buildings — foundations, slabs, columns, walls, girders, and more. Tells you what you need to know about admixtures, reinforcing, and methods of strengthening concrete, plus tips on field mixing, transit mix, pumping, and curing. Explains how to design forms for maximum strength and to prevent blow-outs, form and size slabs, beams, columns and girders, calculate the right size and reinforcing for foundations, figure loads and carrying capacities, design concrete walls, and more. Includes a CD-ROM with a limited version of an engineering software program to help you calculate beam, slab and column size and reinforcement. **256 pages, 8½ x 11, $39.50**